ICME-13 Monographs

Series editor

Gabriele Kaiser, Faculty of Education, Didactics of Mathematics, Universität Hamburg, Hamburg, Germany

Each volume in the series presents state-of-the art research on a particular topic in mathematics education and reflects the international debate as broadly as possible, while also incorporating insights into lesser-known areas of the discussion. Each volume is based on the discussions and presentations during the ICME-13 conference and includes the best papers from one of the ICME-13 Topical Study Groups, Discussion Groups or presentations from the thematic afternoon.

More information about this series at http://www.springer.com/series/15585

Brian Doig · Julian Williams ·
David Swanson · Rita Borromeo Ferri ·
Pat Drake
Editors

Interdisciplinary Mathematics Education

The State of the Art and Beyond

Springer Open

Editors
Brian Doig
Faculty of Arts and Education
Deakin University
Victoria, VIC, Australia

Julian Williams
Ellen Wilkinson Building
The University of Manchester
Manchester, UK

David Swanson
Manchester Institute of Education
The University of Manchester
Manchester, UK

Rita Borromeo Ferri
Institut für Mathematik
Universität Kassel
Kassel, Hessen, Germany

Pat Drake
Victoria University
Melbourne, VIC, Australia

ISSN 2520-8322 ISSN 2520-8330 (electronic)
ICME-13 Monographs
ISBN 978-3-030-11065-9 ISBN 978-3-030-11066-6 (eBook)
https://doi.org/10.1007/978-3-030-11066-6

Library of Congress Control Number: 2018966113

This Springer imprint is published by the registered company Springer Nature Switzerland AG
The registered company address is: Gewerbestrasse 11, 6330 Cham, Switzerland

Contents

1 **Introduction to Interdisciplinary Mathematics Education** 1
Brian Doig and Julian Williams

Part I Conceptualising and Theorising Interdisciplinarity in Research, Policy and Practice
Julian Williams

2 **Introduction** 9
Julian Williams

3 **Theoretical Perspectives on Interdisciplinary Mathematics Education** 13
Julian Williams and Wolff-Michael Roth

4 **Integration from a Commognitive Perspective: An Experience with Mathematics and Music Students** 35
M. Alicia Venegas-Thayer

5 **Challenges and Opportunities for a STEM Interdisciplinary Agenda** 51
Russell Tytler, Gaye Williams, Linda Hobbs and Judy Anderson

Part II Focus on Cross-Cutting Skills: A Glass Half-Full?
Pat Drake

6 **Introduction: A Glass Half Full?** 85
Pat Drake

7 **Developing Mathematical Reasoning Using a STEM Platform** 93
Andrzej Sokolowski

8 **Quantitative Reasoning and Its Rôle in Interdisciplinarity** 113
Robert Mayes

9 **Modelling and Programming of Digital Video: A Source for the Integration of Mathematics, Engineering, and Technology** 135
Carlos A. LópezLeiva, Marios S. Pattichis and Sylvia Celedón-Pattichis

Part III Case Studies in Inter-Disciplinarity: Mathematics as Tool and Mathematics as (Conscious) Generalisation
David Swanson

10 **Introduction** 157
David Swanson

11 **Mathematics in an Interdisciplinary STEM Course (NLT) in The Netherlands** 167
Nelleke den Braber, Jenneke Krüger, Marco Mazereeuw and Wilmad Kuiper

12 **Maths Adds up** 185
Maite Gorriz and Santi Vilches

13 **The Successful Students STEM Project: A Medium Scale Case Study** 209
Linda Hobbs, Brian Doig and Barry Plant

14 **"Draw What You See" Transcending the Mathematics Classroom** 229
Signe E. Kastberg, Rachel Long, Kathleen Lynch-Davis and Beatriz S. D'Ambrosio

15 **Inter-disciplinary Mathematics: Old Wine in New Bottles?** 245
Brian Doig and Wendy Jobling

Part IV Teacher Education and Teacher Development
Rita Borromeo Ferri

16 **Teacher Education and Teacher Development** 259
Rita Borromeo Ferri

17 **Inclusion of Interdisciplinary Approach in the Mathematics Education of Biology Trainee Teachers in Slovakia** 263
Ivana Bobońová, Soňa Čeretková, Anna Tirpáková and Dagmar Markechová

18 **Creating Academic Teacher Scholars in STEM Education by Preparing Preservice Teachers as Researchers** 281
Jennifer Wilhelm and Molly H. Fisher

Part V Conclusion to Interdisciplinary Mathematics Education
Brian Doig

19 Conclusion to Interdisciplinary Mathematics Education 299
Brian Doig and Julian Willams

Chapter 1
Introduction to Interdisciplinary Mathematics Education

Brian Doig and Julian Williams

Abstract The purpose of this chapter is to preface, and introduce, the content of this book, but also to help clarify concepts and terms addressed, set the stage by summarising our previous work, and issue some caveats about our limitations. We will close with a discussion of the mathematics in *Interdisciplinary Mathematics Education* (IdME), which we see as a lacuna in the literature, and even in this book.

Keywords Interdisciplinary · Mathematics · Education · Introduction

1.1 Origins and Context of This Volume

The origins of this book emerged after some of us were invited to lead a Topic Study Group (TSG-22) at the International Conference on Mathematics Education in Hamburg in 2016 (ICME-13). Initially the suggestion was for a topic on *Science, Technology, Engineering and Mathematics* (STEM) education, a topic that has been increasingly prominent in educational policy, and practice, in the last decade. However, we preferred to go to the concept of 'interdisciplinarity' as the focus of interest for several reasons. First, while much STEM-related work does involve interdisciplinarity, much does not—it had emerged as a funding priority, and is often related more to political and economic expediencies, than to educational principles. Second, much STEM work does little to emphasise mathematics, and when it does so, it does not always relate the mathematics to the other disciplines or subjects involved. Third, much good interdisciplinary mathematics involves non-STEM disciplines, and we wanted to include the arts, music, and other disciplines that might be excluded from STEM. Finally, however, we hoped that "Interdisciplinary mathematics education" would include much STEM work, and even most of STEM work, that might be of contemporary interest.

B. Doig
Deakin University, Melbourne, Australia

J. Williams (✉)
University of Manchester, Manchester M13 9PL, UK
e-mail: Julian.williams@manchester.ac.uk

B. Doig et al. (eds.), *Interdisciplinary Mathematics Education*,
ICME-13 Monographs, https://doi.org/10.1007/978-3-030-11066-6_1

1.2 The State of the Art in 2016: What Next?

Prior to the ICME-13 conference, the organisers were invited to produce a *State of the art* in the topic, which was published immediately prior to the conference, and is available freely on-line (see, Williams et al., 2016). This provided the base level of knowledge that all papers in Topic Group 22 built on, and many of the chapters in this book refer to it, so it is worth summarising some of its key points here.

In the *State of the art*, the authors make clear that previous research in the topic suffers from several key problems, or even flaws. First, there is confusion over the key concepts and terms, making it hard for research to become cumulative. Much of the writing in the topic assumes that a discipline equates with a school curriculum subject, and that any form of collaboration, or integration, between subjects is therefore 'interdisciplinary', whereas, we prefer to use the term 'curriculum integration' or 'subject integration' in such cases, unless there is, also, a clear case of interdisciplinarity, i.e. of distinct disciplines working together at some level (more of these 'levels' later). *The State of the art* addressed this concern, and this is further developed in this book, particularly in the first section. However, we will have to face the fact that the term 'discipline' has multiple uses, meaning somewhat different, though overlapping, things inside academia, and outside it in workplaces. In places such as health services, for example, multi-disciplinary teams tend to refer to teams involving distinct *professions*; some of which may arguably have an academic discipline, or two, in their background training, but the academic disciplines involved do not determine the profession. The inter-professional education being promoted there does, however, engage with many of the issues involved in academic 'interdisciplinarity', then, but the concepts involved are different.

Second, the practices involved in schools and tertiary institutions that are quite reasonably described as interdisciplinary, may take a number of forms, and empirical researches into these practices do not often make clear what form they are studying. Again, this makes accumulation of knowledge difficult, and meta-analyses almost impossible. A particular concern comes when these practices are studied empirically, using measurements of learning outcomes; the learning out-comes rarely, in fact, correspond with the interdisciplinary learning outcomes that might have been anticipated or that motivated the practice in the first place. Studies that use traditional test scores in mathematics as an outcome measure, for example, might produce disappointing findings, because the traditional measures are not designed to measure what interdisciplinary practices are designed to develop in learners.

Nevertheless, empirical studies have consistently shown that the raft of school practices, called interdisciplinary, have positive impact in at least one respect—that is the attitudes of teachers and learners to these innovative practices, which usually involve the disciplines being called upon to help learners solve problems in some sort of inquiry classroom practice. It is difficult to extract this dimension, which also is clearly present in the mathematical modelling literature, from the fact of interdisciplinarity as such.

Third, the literature is awash with case study descriptions of practices, usually motivated by enthusiastic practitioners. These are valuable for other practitioners with similar enthusiasms, in helping to prepare them for the possibilities, including opportunities and disappointments, in developing similar practices: there are examples in this book. However, such case studies are doomed to remain local, unless they are accompanied by an analysis of the phenomena of which they are proposed to be a case. Thus, in reviewing this mass of literature, purely descriptive works may make little impact on the field. For this reason, in this book, we have encouraged authors to make a clear statement of the way their work makes a contribution to understandings of policy, or practice, or to the academic literature. Many authors have done so in the conclusion to their chapter.

The *State of the art* in 2016, then, does proceed to offer some solutions to these problems and to this, many of the contributions in the chapters in this book can now be added. In the concluding chapter we will summarise the main additions to the *State of the art* now, in 2018! But here in the introduction, we will prepare you for this, a little, with an advanced organiser.

Most importantly, we suggest, that for the purposes of research, an academic discipline (such as mathematics) be defined in relation to other disciplines which it has developed a division of labour. It is then understood as a cultural-historical formation of practices (such as mathematical practices) that require teaching (hence, disciples and discipline) as a body of knowledge, discourse and skill. Clearly these are at least partly organised in specialised institutions, whose rôle is to build, and, or, develop the discipline as a distinct practice, as well as to protect its particular values, epistemology, communities, and institutions. To this, it is crucial to add, the emotions. All of this means framing disciplinary and interdisciplinary mathematics as a cultural practice, which is inspired by Vygotskian thinking, and Activity Theory.

Implicit in this, however, is the possibility—maybe the necessity—of power hierarchies, alienation, privilege, exclusion, *et cetera*. Indeed the separation of the cultivation of a discipline in (often privileged) academic institutions separated from its practice in everyday life activity almost defines it as alienating and alienated. These features are discussed well, in Bourdieu's and Foucault's *oeuvres* (see again, Williams & Roth, this volume). Indeed, it may be argued, that it is important to understand this sociology in order to challenge unfortunate, or oppressive, disciplinary practices. The formulation of practice as the dialectic of *habitus* and field (i.e., a field of power), or its formulation as a discourse, leads to distinct analyses of power, but they converge on the notion that resistance or challenges to this power require reflexivity, or a meta-awareness of the field, and its discourse.

A major topic to come, then, is one that has hardly been researched in the context of interdisciplinarity, and that is meta-cognition, including meta-knowledge of the disciplines.

1.3 The Sections and Chapters in the Book

The contributions to the ICME-13 conference Topic Study Group included many presentations, papers, and posters, and authors were invited to consider making a full length chapter for this book: their topics fell fairly naturally into four sections. The first section, sub-edited by Julian Williams, on *Theorising and conceptualising IdME for policy and practice*, comprises three chapters. In the introductory preface to this section, Williams summarises these as being theoretical and conceptual, clarifying the distinctions mentioned above and: developing notions of different kinds of discipline and interdisciplinarity (Williams & Roth, this volume); adding the commognitive perspective of Sfard in a case study of mathematics and music (Venegas-Thayer, this volume); and developing understandings of educational policy and practices in a diversity of cases (Tytler et al., this volume).

The section, sub-edited by Pat Drake, a *Focus on cross-cutting skills*, focuses on the state of practice, as experienced by the chapter authors. Drake notes two important, but, perhaps, not surprising, things. The first is that the chapter authors believe, strongly, that it is important to make a coherent contribution to the field. This is because, as the authors suggest, there are as many definitions of interdisciplinarity as there are commentators. This would place interdisciplinary mathematics as a being in a gestational period, not, as yet, a mature field of research or practice. Whether this state of affairs continues, of course, may well be influenced by the present volume, depending on how readers subscribe to the views expressed in this section. Furthermore, if theorists, and researchers, disagree, or at least, do not have a sufficient degree of coherence, practitioners will be left to determine what is possible in school contexts, which may lead, permanently, to interdisciplarity mathematics education being some sort of hydra, rather than a singular beast.

The second thing that struck Drake, was that the chapter authors had cited, completely different sources, indicating that no one text stands out as defining the field, or, as Drake puts it 'clearly there are currently no seminal texts, no shared body of work, on which to build our understanding'. This, of course, would further indicate that, at present, we are dealing with a hydra.

This, as Drake says, simply emphasises the need to consider the importance of developing 'a systematic and international review of the field', that would support research endeavours with clear, and universally acceptable, foci.

The third section, sub-edited by David Swanson, on *Inter-disciplinarity: Case studies in inter-disciplinarity (mathematics as tool and mathematics as (conscious) generalisation)* looks at, as the title suggests, some illustrative case studies of inter-disciplinary mathematics education.

In his introductory remarks for this section, Swanson makes a strong argument that these chapters can help the reader gain an 'understanding [that] helps us see how mathematics can be of use in interdisciplinary work beyond its rôle as a tool'. How mathematics has been portrayed, and used, only as a tool, is suggested by a quote from Freudenthal, that claims that school mathematics, in general, has not only cut the bonds between mathematics and reality, but also 'between mathematics and itself'. Therefore, interdisciplinarity mathematics, Swanson suggests, needs to start with the bonding of 'mathematics, extra-mathematical reality, *and* between mathematics and itself' in order to create a form of mathematics that can 'dominate and then mathematics could share its full richness within interdisciplinary work', and in so doing, shed its reputation as a mere tool for use with other disciplines.

Swanson found that the chapters in this section emphasised that 'engaging with real world problems (in project work, interdisciplinary problems, or art) brings great benefits for the learning of mathematics'. However, is this really interdisciplinary, or is such project work simply disparate subjects being used as a collection of tools?

The fourth section, sub-edited by Rita Borromeo Ferri, *Teacher education and teacher development* offers 'insight into good teacher education approaches for interdisciplinary teaching and learning'. Borromeo Ferri states that the chapters, in this section, underscore how critical it is 'to make the effects of… interdisciplinary teaching explicit and transparent for colleagues', which is a point not necessarily taken up by either researchers, or practitioners. Clarity of the nature of one's practice could be a major step in assisting the general understanding of interdisciplinary mathematics. However, spreading the word is not sufficient: there needs to be opportunities for researchers, and practitioners, to explain and debate how, and why, a particular approach *is* 'ínterdisciplinary mathematics'. However, Borromeo Ferri claims that the chapters in this section 'show that [while] there is a lot of progress in STEM teacher education', there remains a need for teachers who are educated in interdisciplinary mathematics teaching. While this is an obvious point to make, she also adds, that 'we need teachers, who are open minded enough not to see only their own favoured subject or discipline, but who like to connect several disciplines, [and who] discuss their links with colleagues, create ideas and make interdisciplinary teaching and learning lively and motivating for the students'. Whether this is obtainable in the near future, most likely depends upon the influence that these chapters, and others like them, can exert on the teaching profession.

In comparison to previous work such as the state of the art outlined in Williams et al. 2016, this volume builds on and expands the knowledge base of concepts, theories, examples, and studies relevant to this new field. One particular thread that arises again and again from research into interdisciplinarity is the meta- cognition of the discipline of mathematics, and its relations with other disciplines and knowledges. We will pick up this thread again in our concluding discussion, when we finally conclude by asking 'what is still to be done?'

Reference

Williams, J., Roth, W.-M., Swanson, D., Doig, B., Groves, S., Omuvwie, M., et al. (2016). *Interdisciplinary mathematics education: State of the art*. Cham: Springer. https://doi.org/10.1007/978-3-319-42267-1.

Part I
Conceptualising and Theorising Interdisciplinarity in Research, Policy and Practice

Julian Williams

Chapter 2
Introduction

Julian Williams

Abstract This section introduces interdisciplinarity in theory, and its conceptualization for policy and practice, with a view to developing the research in this subfield as a cumulative, scientific enterprise. The three chapters are outlined. Then I appeal to theory, policy and practice to re-think the notion of school discipline, and to unleash the learner, teachers, and the schools from discipline.

Keywords Interdisciplinary · Theory · Mathematics

The purpose of this section is (i) to summarise the state of the art, including the three chapters in this section, and (ii) to go beyond and indicate some directions of travel for research in the future, in the theorisation and conceptualisations of Inter-disciplinary Mathematics Education (IdME) for research, policy and practice.

But first, something must be said about the phrase 'for research, policy, and practice'. Any review or summary of literature worth its salt must have a direction or object, and here we have three! The aim 'for research' must be to try to establish clarity about concepts, terms and perhaps epistemology and methods so that workers across the field can speak to each other and accumulate knowledge. In addressing the concerns of interested, especially professional, researchers one might expect a modicum of investment of effort and time to grasp the complexity of the issues. The aim 'for policy' is to provide research outcomes in a way that informs policy concerns, such as how to prioritise, what to legislate, and fund. The aim for practitioners might be more illustrative, offering principles that are enriched by cases, and that engage with reflective accounts of practitioners' experiences, both inspiring and troubled. To achieve all three at once is perhaps impossible and different sections and chapters will be more weighted to one or another aim. But let's 'be realistic, and demand the impossible'.

This section's first chapter by Julian Williams and Wolff-Michael Roth develops a sociocultural, cultural historical account of 'disciplines' that begins in the social division of labour, evolves through various social formations like the guilds, and leads to the modern alienation of academic scholarship from productive labour in schoolified

---

J. Williams (✉)
University of Manchester, Manchester M13 9PL, UK
e-mail: julian.williams@manchester.ac.uk

B. Doig et al. (eds.), *Interdisciplinary Mathematics Education*,
ICME-13 Monographs, https://doi.org/10.1007/978-3-030-11066-6_2

institutions whose curriculum is essentially specialised, academic knowledge, and whose product is a stratified, 'classed' potential workforce. This alienation of the disciplines from productive activity (i.e. labour, work, even research) then poses the challenge of meaningful learning of both disciplinarity and interdisciplinarity, which the chapter helps clarify in its various categories.

This chapter also reflects on the meaning of multidisciplinarity outside the academy, in workplaces: this is an important context because here (e.g. in multidisciplinary teams in work places) teams of professionals are obliged to work together effectively even in order to execute their own profession respectably, and this involves subduing (maybe one can say 'sublating' or superseding) their own disciplinary priorities in favour of prioritising the joint objective (e.g. the health of the patient, or the welfare of the child, etc.).

This kind of interdisciplinary working highlights the need for professionals to not only 'know their own stuff' but also to have some minimal understanding of how other professions operate and how to interface with them (Wenger refers to this as 'knowledgeability' in landscapes of practice). 'Knowing about' a discipline in this way is referred to as part of meta-disciplinary knowledge of a discipline, not something that schools teach explicitly, though perhaps learnt as part of an implicit, hidden curriculum. This 'knowing about' certainly involves, for instance in mathematics, knowing when to use a discipline and when NOT to. Unfortunately, traditional curricula do not deal with this concern, or not very well: as a consequence learners may 'learn' that mathematics is really pretty useless, something one does at school, with little purpose or relevance to ones concerns and interests. The chapter then finishes with an argument for unleashing the learner from the disciplines in problem solving, which can only arise from (i) knowing a bit about and having some know-how within the discipline, and (ii) knowing when it is appropriate and when not. The authors use the phrase 'becoming undisciplined' accordingly.

The second chapter in this section by Alicia Venegas adopts Sfard's Vygotskyan perspective on discourse and knowledge called 'commognition', which holds that thinking, knowing and knowledge are to be understood as discursive 'communication'—with others, and then also with oneself. In this perspective, a 'discipline' is a discourse, or has a discursive structure (defined by keywords, signs, routines, and endorsed narratives of truth); and working across disciplines involves a sort of integration of the discourses involved, e.g. by translation between two different language structures. In order to illustrate this concept of interdisciplinary integration, discourse from a team of mathematics students and music students is analysed. The team is collectively engaged in making stochastic music, where a set of random variables with appropriate probability distributions decides the number of fingers, notes, pitch, and octave of a random unit of sound. The need to speak of this set as a unit led, through gestures and exploratory talk, to a special, invented term "baggie"—meaning not just a musical sound but also, simultaneously, a mathematical set of random numbers. Importantly, when the mathematicians talk of a baggie they are thinking of its mathematics, but the musicians think of it as a musical chord (actually there is a moment where it becomes natural and desirable to work on two 'baggies' to represent the left and right piano hand).

This line of research has important conceptual elements that draw on Vygotskyan theories of language and communication, but also resonates with the third generation of activity theory discussed in Williams and Roth's chapter: clearly the baggie is a boundary object that manifests the contradictions between the two communities meanings that are held to be developmental. To Vygotsky one might add Bakhtin. The discourses of Sfard might usefully be thought of as language genres in the Bakhtinian sense. Then one would note the significance of the national language which services these different genres and in particular support communication in multiple genres and discourses (hence 'heteroglossia'). The 'boundary object' called a baggie thereby becomes a hybrid word, animated by two distinct discourse genres. In this chapter's case, communication across the discourses is facilitated by the common language of Spanish, but also notably that of gesture.

In the third chapter in this section, Russell Tytler and colleagues discuss integration in a different sense and context, using cases of interdisciplinary projects in schools involving Science, Technology, Engineering and Mathematics (STEM). The lessons from these practical experiences are drawn out in the chapter's coda, and align well with the previous literature (e.g., Williams et al., 2016) in terms of the affective outcomes, and even of the dangers of mathematics becoming routinised into a mere 'tool' for the problem solving by science and technology. The cases show how mathematical creativity could be a significant part of the problem solving, and thus maintain the 'epistemic integrity' of the subject within motivating interdisciplinary contexts.

But also, there is a substantial emphasis here on caveats regarding the amount of planning work, support from the institutions' management, and even outside expertise needed. At the heart of these practical problems, I would argue, is the fact that schools are not really built for interdisciplinary curricula and pedagogies (and I even mean 'built' in the architectural sense, as well as metaphorically in the policy, professional structure, and curriculum and assessment imperatives in the system). This is reminiscent of the observation of a female engineer that the boots offered to visitors on the building site were always too big: in general, the infrastructure emerges historically in ways that are not fit for the purpose of today's tasks; and we are, as Marx observed, condemned to make history with the tools bequeathed us by history. Thus, the institution requires that every project has to be 'open' to student creativity and inquiry (engaging motivation), and yet planning has to ensure that there will be an element of new disciplinary (including mathematics) learning of the appropriate kinds and levels laid down in a pre-historic curriculum (and pedagogy, and …). This has been called 'pushing water uphill' (that is, not impossible but eventually exhausting, according to Sisyphus).

Imagine an education where there is space for teams of students to simply identify problems of interest and then work on them, with a raft of expertise on hand to help where necessary. Then, how would the curriculum be specified, taught and assessed, and how would teaching or expertise be made available in the right forms and places, and at the right times? I think this is a useful question, as it provokes imagination of educational activity led by inquiry and engaging interdisciplinarity, but not artificially so. Mathematics and other disciplines would have to fight for space in a busy field of knowledge and know-how. The contradictions between alienated disciplines would still be present, but the institutional order would be upside down. The outcomes of the traditional curriculum are strata of people classified by multiple grades in separate subjects and disciplines, but in the upside down university and school, the outcomes might be problems solved, and problem solvers educated by and with a variety of productions, inscribed on their CV, although perhaps with disparate skills and knowledge.

In a previous 'vision' project some of us explored this possibility for the Royal Society of London: this vision remains available online five years later for the interested googler, but apparently untouched and unremarked upon by any serious educational policy maker that I am aware of. Well: why would anyone be surprised?

Reference

Williams, J., Roth, W.-M., Swanson, D., Doig, B., Groves, S., Omuvwie, M., et al. (2016). *Interdisciplinary mathematics education: State of the art*. Cham: Springer. https://doi.org/10.1007/978-3-319-42267-1.

Chapter 3
Theoretical Perspectives on Interdisciplinary Mathematics Education

Julian Williams and Wolff-Michael Roth

Abstract In this chapter, we develop in broad strokes the concept and history of the 'disciplines', a prerequisite for understanding disciplinary and interdisciplinary activity, since activity is always mediated by the cultural artefacts history leaves us. We develop the social and cultural theories of activity, practice, and discourse to offer further insights into both academic and professional 'disciplines', and their interrelationships, both in the academy, and in practical, joint, 'interdisciplinary' activity in everyday, workplace and professional life. The aim is to provide the foundations of a comprehensive theory for researchers of interdisciplinary activity. We build the analysis first of all on classical activity theory and modern developments in this tradition (a) of Vygotsky's group and their Western interpreters, and (b) of those inspired by Bakhtin who have particularly developed multivoicedness and hybridity in dialogism. We additionally draw on Bourdieu and Foucault to consider the nature of the power structures in the disciplinary fields and discourses respectively, and how they might be resisted. We argue for a new conceptualisation of meta-disciplinary mathematics education that is a requirement of a critical mathematics education, concluding that meta-knowledge of disciplinarity is necessary for negating and becoming, to some extent, free from the discipline. We reflect on the adequacy of this theoretical battery, and its proposed synthesis for researchers in the field.

Keywords Interdisciplinary · Mathematics · Theory · CHAT · Bakhtin · Bourdieu · Foucault

---

J. Williams (✉)
University of Manchester, Manchester M13 9PL, UK
e-mail: julian.williams@manchester.ac.uk

W.-M. Roth
University of Victoria, Victoria, Canada
e-mail: wolffmichael.roth@gmail.com

B. Doig et al. (eds.), *Interdisciplinary Mathematics Education*,
ICME-13 Monographs, https://doi.org/10.1007/978-3-030-11066-6_3

3.1 Introduction to Interdisciplinarity

The problem of interdisciplinary mathematics education requires a conceptualisation of 'discipline' or 'disciplinarity'. But first, assuming an everyday, intuitive notion of discipline, we reflect on the issue of 'inter' disciplinarity: in fact, interdisciplinarity is often understood as a multifaceted but partially nested system of concepts, where different forms of inquiry are situated, at one or another level of complexity of the inquiry process, as follows (see a more extensive discussion of this in Williams et al., 2016) (Fig. 3.1).

Here it is suggested that 'interdisciplinary' mathematics involves various sorts of conjunction of mathematics with other knowledge in problem solving and inquiry. This 'other' knowledge is generally outside of mathematics, whether this involves one or more other disciplines (e.g. when mathematics is used as a tool within a science, geography or social sciences project) or just extra-mathematical, even 'everyday' knowledge (as in mathematical modelling of traffic flows perhaps, where only some everyday knowledge is needed while the mathematics might be very deep). As the relationship between mathematics and other disciplines becomes more interconnected, a genuine 'inter' disciplinarity emerges, when mathematics interacts with other disciplines to become something new and different (e.g. when mathematics, statistics and sociology become a new, hybrid 'quantitative reasoning', or in mathematical-physics, and mathematical-biology).

'Trans-disciplinarity' usually implies transcendence due to some sort of subsuming of the disciplines within a joint problem solving enterprise, and here the disciplines are not necessarily consciously marked, and as such, may almost seem to disappear. Essentially this is because the focus of attention is on the problem at hand, and disciplines merely provide tools for achieving a solution. In statistics, for example, the properties of distributions are of interest to mathematicians, whereas social scientists may focus on the mobilisation of the data for the purpose of comparing population characteristics. In school STEM projects, mathematics often disappears into the science and technology involved, and in fact the disappearance of mathematics quite generally in black boxes has long been noted (Williams & Wake, 2007). The fact is that when the motive of activity is to solve some real problem, some disciplines may prove irrelevant to, or even get in the way of, finding a workable or effective solution. When engaged in the process of crossing the road or overtaking a car, it may not be helpful in the moment to reflect on the kinematics involved, though some modelling of traffic may be helpful to road planners in positioning road signs and crossing facilities, or in determining speed limits.

Fig. 3.1 The spectrum of interdisciplinarity in problem solving, after Williams et al. (2016)

Mono-disciplinarity →
multi-disciplinarity →
inter-disciplinarity →
trans-disciplinarity →
meta-disciplinarity →

Finally on this spectrum, we add meta-disciplinarity, involving awareness of the nature of the discipline or disciplines involved. This becomes relevant when one becomes aware of the root disciplines—including mathematics—in their relation and difference within inquiry, e.g. when the nature of 'using evidence' in history and in science becomes contrasted. Thereby the epistemic qualities of the disciplines become clearer, but this is the stage at which conscious, theoretical control of the disciplines becomes possible: one can finally say, for instance, 'this is a problem or task or situation where mathematics will likely be relevant (or not relevant)' and even give reasons for this decision.

This kind of meta-knowledge can emerge from reflection on the relationship of mathematics or other disciplines with other knowledge at any point on the above spectrum, and so perhaps should be placed above the axis from 'mono' to 'trans'. We will argue in this chapter that it prompts a new 'cycle' of the spectrum, where the next cycle emerges at a higher level potentially integrated with meta-knowing.

But the whole idea of the spectrum of complexity itself is perhaps incomplete, as it is not quite clear how an academic 'discipline' and 'other knowledge' should be understood: we will revisit this spectrum again after exploring these notions more carefully. In our conclusion, we will argue that it may be desirable to think of a beyond-disciplinarity which is not only 'meta' in the above sense, but which we will term 'knowingly un-disciplined', i.e. to some extent freed from the disciplines that bind problem-solving and inquiry to disciplinary norms and their limits (Williams, 2016).

3.2 Professional Disciplines

But now we come to the notion of 'disciplinarity' in the professional sense, in the world outside of academic 'sciences' proper: for instance, we may speak of multi-disciplinary teams in and around the health service. Here the disciplines may appear superficially in different job titles and remits, such as nurse, teacher, general practitioner, and social worker, but successful activity often involves effective inter-professional teamworking, often described as 'multidisciplinary'. In the out-of-school context one sees many of the same issues arising in joint work as one does with interdisciplinary work in academe and science: but now team work, professional or disciplinary 'identity', and division of labour are absolutely essential and must somehow be subsumed in the holistic interest of the motive, e.g. the 'health of the patient'. Thereby, each professional 'discipline' then has some sort of professional identity at stake, but must also prove itself as efficacious to the larger good, in the 'joint enterprise' or activity. Each discipline, even to justify itself, has to allow itself to be subsumed into an integrative whole, which has its own dynamic that is likely to differ from the dynamic of each contributing discipline. When more than one profession is involved, then some practical awareness of the possibilities for relations between the professions is required, which Wenger (1998) calls 'knowledgeability' in the landscape of communities of practice, and which Edwards (2017) has called 'common knowledge' that gives rise to relational expertise and hence 'relational agency'. The point here is that teamwork involving distinct disciplines demands of professionals not just that agents know about the other teamworkers' disciplines, but that they practise their relationships collaboratively with other disciplines in effective ways in the interest of their shared aim or objective (Edwards, 2017).

In addition, professional disciplines also often have their scholarly academic and practical 'knowledge bases'—though their professionalism may be defined more often by the practical competences and membership of a professional association than by their formal academic accreditation or disciplinary qualification as such. Indeed many of these professional disciplines have spawned specialist schools in academia, as they demand academic qualifications and accreditation as a minimum for entry to their profession: schools of engineering, medicine, and now nursing, social work, and even film, computer games and so on are or are becoming commonplace in universities. Indeed, some have been present from the foundations of many, even the medieval, universities (e.g. theology, medicine, and teaching). In the next sections then we seek to illuminate both *academic* and *professional* disciplines in a general theory or conceptual framework of disciplinarity, within a historical, social context.

3.3 Disciplinarity in Sociocultural Activity Theory

The journey over the terrain of interdisciplinarity must begin by seeking to understand why disciplines in our general sense arose and continue to flourish, even producing new sub- or hybrid-disciplines. We need to see how they work both separately and together to service social functions. Only then can we understand the difficulties and constraints—but also the opportunities—that interdisciplinary work poses.

Disciplinarity is both (a) a social phenomenon caused by increasing specialisation and differentiation of labour involving social, material and discursive practices, and (b) a form of discourse making the specialisation 'thematic' (located in a coherent body of transmittable or 'teachable' knowledge). Although this division of labour preceded the birth of formal teaching, these two aspects of disciplinarity have become mutually reinforcing. The term itself derives from the Latin *disciplina*, meaning teaching, instruction, training, branch of study, philosophical school, monastic rule, and chastisement. By Chaucer's time 'discipline' was used to refer to branches of knowledge, especially to medicine, law, and theology, the 'higher faculties' of the university. In the sociology of knowledge, the origin of science has been situated in the religious forms of life, with associated rites, bodily discipline, and asceticism; modern science, which in certain respects negated religion, nevertheless is characterised by rules and norms derived from the religious practices and discourses it largely has replaced (and as such harbours gender, racial, and class historicity).

In any sociological account, disciplinarity is treated as a social phenomenon. The smallest unit of analysis for any *specifically human* social phenomenon has to be one that has the key characteristics of society we wish to analyse. One such unit is 'productive labour activity', involving the production of things for consumption, i.e. meeting human needs. Productive activity engaging with a particular, distinctive, material and discursive practice, together with its consumable products, can be seen as defining the functions of any discipline, and is the means to understand a discipline's abiding meaning and structure. Thus, the discipline of mathematics or medicine should be understood in its functionality, whether in producing effective transport or healthy humans.

In this view, disciplinarity did not exist from the beginning of humankind, but came into being as 'disciplined activity', to meet some specialised need. The many different forms of production that exist today historically have emerged as a result of increasing division of labour and specialisation, making production more efficient or its products more effective (or meeting new, anticipated needs). But in some specialties an infrastructure of knowledge requirements grew, such that teaching in some form became functional and even necessary: and this teaching begins to characterise the specialised activity of an emerging discipline. The old-timers in a profession become the teachers, and their nurturing of a body of knowledge becomes a key component in the production and reproduction of the discipline. This nurturing of knowledge through teaching, when it becomes formalised in a curriculum and school, becomes disciplinary activity of a different kind from the specialised activity that gave birth to it, and typically becomes alienated from productive labour as such in academies and schools. In previous work, for instance, we pointed out how (and why) the practice of graphing takes quite distinct forms in schools and workplaces. This is because of the way assessment shapes school activity, which must have the appearance of equity (Williams, Wake, & Boreham, 2001; Williams & Wake, 2007).

Any productive activity generally involves collective, joint labour. It may be characterised by a unity of a number of moments, including the *subject(s)* dialectically engaged with the *object (or 'object-motive')* of activity. This dialectic is significantly mediated by the whole historically evolved, social and material system of production, involving means of production such as tools and signs, but also conventions, and the division of labour between subjects governed by rules and regulations, see Engeström's schema (1987) adapted in Fig. 3.2.

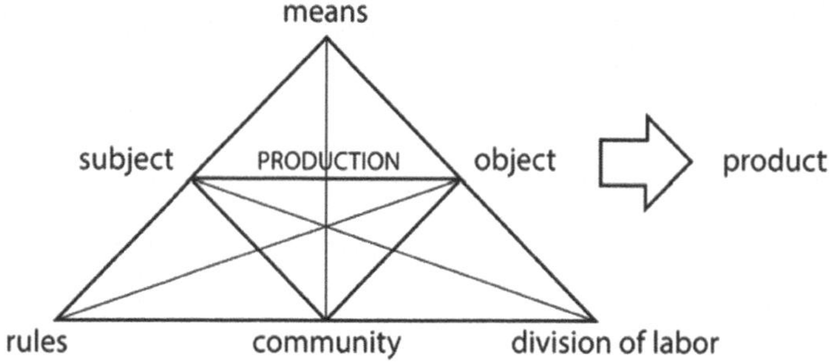

Fig. 3.2 Cultural-historical activity theoretic schema of an activity system. From Engeström (1987), after Leontiev (1978, 1981)

Leontiev exemplified his concept of 'activity system' (defined as the whole 'ensemble of social relations' as Marx put it) in the case of the pre-historic primal, collective hunt. The hunter(s) are subject(s) jointly engaged in hunting the game (the material 'object' of the hunt) in order to meet their needs (consumable food, feathers/fur for making clothes, etc.)—thus the hunting activity engages a community in transforming its object (the free, live animal) into a serviceable outcome (food, clothing). The 'motive' of the hunt is precisely this envisaged outcome and Leontiev says this 'object-motive' is what defines the activity as such, and what makes it functional for the community. The hunt as an 'activity' is collective, but is constituted by a number of coordinated, observable individual goal-driven 'actions' (the tracking or frightening of the prey into the open, the transport back to camp, the preparation, cooking, and feasting) mediated by certain tools (weapons, beaters) signs (calls, speech) and norms (who does what and how—the killing, cooking, sharing). There may be a division of labour, with some flushing the game out of their nests/hides while others prepare for the kill. The rules of the community may indicate that certain roles in the hunt have prestige, command special value and so on, but in primitive societies the division of labour is generally rudimentary, and in comparison to modern labour the whole activity and its object is relatively visible and maybe conscious.

In complex modern labour activity, often the activity as a whole is not so visible, and one only becomes conscious of it through reflection or analysis, even when one is aware of one's own role and actions that are part of the whole enterprise. Leontiev expressed interest in the moment when one becomes conscious of the motive of the activity as a whole (e.g. why we should study history) or of the operational level and the functioning of tools in our actions (as in breakdown moments when one becomes aware of the significance of mediating tools like computers or signs like language).

In understanding the functionality of a discipline like mathematics in productive activity then one may see it in the conceptual instruments (or even embedded in the physical instruments such as diagrams or measurement tools) that afford productive activity *with* the discipline (it is only in mathematics per se that the mathematics itself becomes the object of the activity). But this then is reflected in all the other moments: the subjectivity of the subject, who might be disposed (or not) to use mathematical means and instruments; their relation to other subjects involved in the activity who might be mathematicians or other specialists; the community or communities involved in the production, or those who might be called upon to engage in some specialised way, and so on. Thus to understand mathematics as a discipline one must understand its function in, and as, labour activity, and in any of the moments of the activity system (Fig. 3.3).

In Leontiev's account, the directedness of activity correlates with affect (emotions), which constitutes a form of consciousness about the current status of the activity and the satisfaction of a need that may arise from the outcome. Whereas in early human activity, the meat was used to satisfy the dietary needs of the hunting party, the fulfilment of a need may be less apparent in modern productive activities, where workers sell their labour for a wage that is used to satisfy a diversity of needs (housing, food, clothing, or leisure). This translates in learning and education to

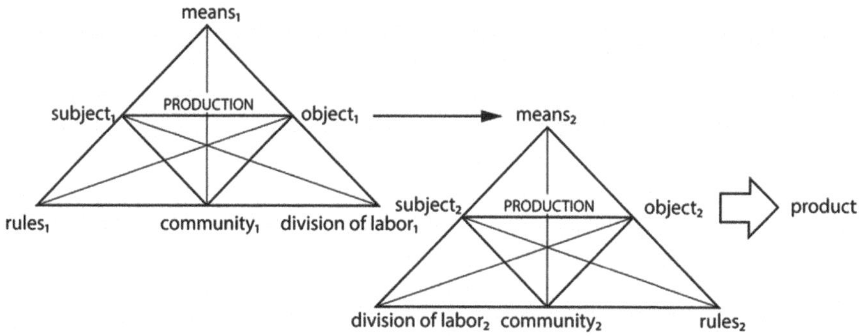

Fig. 3.3 In this arrangement of activity systems, the first activity system has as object/motive the tools required in the second activity system (e.g. industry making the weapons but the military applies it to wage war)

learning outcomes, grades and accreditation needed for progression, social acceptability/respectability, or even a career. Because an activity is a whole unit, every action contributing to its accomplishment also is permeated with this affect.

The triangular schema for cultural historical activity theory (Fig. 3.2) has its weaknesses—as any simple model must. In particular it misses an important element in Leontiev's account of an activity system, in that it fails to distinguish between the individual subject's conscious goal and the social motive of the collective activity as a whole. Another weakness is the static representation of the schema: Leontiev conceived of the activity from the beginning to the end as a whole unit so that every part (subject, object, etc.) is changing *within* this unit. This springs from the problem of schematically representing a dialectic in a diagram, and dialectical relations with line segments. Thus, the subject-object relation should be understood as a dialectical one in which both are moments of change and development, in which each is mediated by the other; so, the object in its current state and the 'ideal' image of the final product together form the object-motive of the activity. Thus, for example, a builder has all materials and tools at hand and the 'ideal' image of the finished house. The object is both the raw material being transformed into a product, but also, in its 'ideal' form, is the motive, i.e. the envisaged outcome of the acting subject. Similarly, the subject is being transformed through activity (one may call this learning, or in some cases development), reflecting their developing actions on the object and relations with other subjects engaged in joint activity.

'Disciplined' mathematical activity may thus be of two very different kinds, according to whether mathematics is only the instrument of labour or also its object—(e.g. for creating new mathematical knowledge). First, an activity may involve labour that benefits from a particular discipline, such as when the performance of a task whose object (e.g. costing a project) benefits from some mathematical disciplined knowledge and actions. Second, activity may have mathematics as the object of activity itself, such as when a theorem is to be proved, or a technique learned. In the first case the mathematics may be instrumental, providing tools for the actions but not their object; while in the second case the object and its product may be mathematics (and the instruments may also be mathematical).

Most importantly for learning, subjects' activity also is dialectically both motivated by, and causative of, the subjects' *consciousness* and *personality* (Leontiev, 1978). In 'disciplined' activity, subjects engage with each other in joint work which obeys disciplinary rules, and thereby transform themselves into disciplined subjects, who may learn and even develop disciplined personalities and identities. An important aspect of the learning of subjects in activity is the contradictory forms of awareness and subjectivities of those involved: thus a subject engaging in joint activity may be confronted by new understandings, practices, and motives, leading to development of consciousness and personality (by 'personality' Leontiev means the structure of one's whole being or life-activity, including the person's character and disposition to engage in certain activities and systems).

Finally, it is impossible to understand the relationship between discipline and institutions if we fail to acknowledge their basis in productive activity and its historically produced mediating conditions, which explain power relations and oppression for instance (Bourdieu, 2000). However, we need to go further in this analysis, contemplating what is involved in the work of the discipline (i.e. its learning and teaching) and how this relates to the productive capacity and thus to the production system outlined above. In the next section we introduce the separation or alienation of the discipline from productive labour as such historically.

3.4 History of the Disciplinary Nature of Human Praxis

Understanding the cultural historical legacy that is entailed in our 'disciplines' may help us to understand the nature of the disciplines themselves. But it also may help us understand why inter-disciplinary work can be difficult, confronting certain sorts of obstacles, power structures, and questions of identity, and differences in understandings of knowledge, discourse and practice.

In classical sociological approaches in the Eurocentric tradition, formal notions of discipline and formal aggregations around particular practices are said to have emerged at the beginning of the Middle Ages, but their Western origin dates back at least to ancient Greece with the emergence of industries besides agriculture (Durkheim, 1893), involving inter-city, inter-state and even international divisions of labour and trade. Discipline as such requires a form of corporation in an institutional form for an aggregate of people, but this institutional form does not in itself constitute a discipline. During the Roman Empire, the different trades came to be treated as entities with particular functions in the public service, the charge and responsibility for which lay with the corporation. Because the service was imposed, requiring state sanctions to maintain it, the corporations ceased to exist with the end of the empire. In the European context, they were reborn in virtually all societies during the 11th and 12th centuries, when tradespeople felt the need to unite, forming the first confraternities.

The confraternities and the guilds they gave rise to, as authorities regulating the practices of their members ('rules' in Fig. 3.2), can be seen as the first organisational structures that exert themselves as forces on the formation of the durable dispositions of its members. Such regulation occurs 'through all the constraints and disciplines that [the organisational structure] imposes uniformly on all agents' (Bourdieu, 2000, p. 175). In the European context, the training of traditional artisans began with *apprenticeship*, which ended when aspiring individuals became *journeymen* upon successful completion of a specific piece of work in and with which they exhibited specific skills. As journeymen, they travelled and worked in different locales until ready to complete a 'Meisterstück' [literally a *masterpiece*, a piece of work to qualify as a master craftsman]' to be judged by members of the guild. Through the piece of work, journeymen exhibited mastery of the means of production (Fig. 3.2) and the form of consciousness required for the transformation of objects into a craft-specific product. If successful, they became *master craftsman* and obtained the right to have their own shop, train apprentices, and employ journeymen. The old forms of reproduction were reborn in the division of training and work, cross cut by another division of theory and praxis: the former occurring in (vocational) school and college, the latter as practical apprenticeship or experiential learning. Even the designation of 'masters' found a new life in the 'Masters degree', and the trade certificates mutated into high school and college/university diplomas.

The increasing division of labour partially is the result of the increasingly specialised knowledge required to do a particular job. 'The production of ideas, of conceptions, of consciousness, is at first directly interwoven with the material activity and the material intercourse of men (sic)—the language of real life' (Marx & Engels, 1974, p. 47). This same progressive division of labour also split theory and practice, the former often being taught in schools, the latter on the job. Indeed, 'division of labour only becomes truly such from the moment when a division of material and mental labour appears' (p. 51).

In the history of intellectual (theoretical) disciplines, 'the specificity of the scientific field stems from the fact that the competitors agree on the principles of verification of conformity to the "real", common methods for validating theses and hypotheses' (Bourdieu, 2000, p. 113). Numerous case studies show how new disciplines or non-disciplinary fields—penology, education, nursing, midwifery, biology, or psychiatry—are tied to: specific, shared discourses and practices; economies of concepts; supporting institutions; conditions and procedures of (social) inclusion and exclusion; transmission and training; relations to law, labour, and morality; and (disciplinary) practices or technologies of surveillance, government, and control (Foucault, 1970, 1978, 1988).

Foucault's archaeological, genealogical and critical studies also reveal who controls existing discourses and how these constitute the very boundaries of any new discipline. As a result, a focus on 'disciplinary boundaries' rather than 'discipline' can help reveal an understanding of the phenomenon as a combination of internal and external social processes. These boundaries are revealed (a) in the relations between distinct disciplines, and (b) in the relation between disciplinary practices and labour activity in which disciplinarity is subsumed.

The academic, scholastic disciplines have their Western origin in the medieval divisions of the trivium (grammar, rhetoric, logic) and quadrivium (arithmetic, geometry, astronomy, music) that lasted to early modernity (d'Ambrosio, 1990). The sciences originated in philosophy, 'which fragmented itself into a multitude of special disciplines of which each has its object, its method, its mind' (Durkheim, 1893, p. 2). The objects of inquiry and the principles on which they are based historically were re-ordered towards the end of the 18th century and the arrival of mathematisation. Before Kant's critique of reason, representations were inherently linked. With *mathesis*—i.e. the systematising practices establishing the order of things—an epistemological differentiation occurred, according to archaeological and genealogical analyses, into a field of 'a priori sciences, pure formal sciences, deductive sciences based on logic and mathematics' and a field of 'a posteriori sciences, empirical sciences, which employ the deductive forms only in fragments and in strictly localised regions' (Foucault, 1970, p. 245).

In sum, a discipline functions as 'a system of control in the production of *discourse*, fixing its limits through the action of an identity taking the form of a permanent reactivation of the rules' (Foucault, 1972, p. 224). One cannot speak 'the truth' outside of such a system, as can be seen in the case of 19th century biology, where the statements of Gregor Mendel about heredity made no sense to contemporaries. It was only after a complete shift in the disciplinary discourse of biology itself that Mendel's statements, its objects and discourse, were recognised as true. That is, one can 'only be in the true … if one obeyed the rules of some discursive "policy" which would have to be reactivated every time one spoke' (p. 224). In this analysis, (disciplinary) forms of discourse, though also an opportunity, first of all need to be thought of as constraint. This constraint arises in part from the acceptable forms of representations and the associated practices that both constitute and distinguish the discipline and its boundaries (e.g. see Lynch, 1985).

3.5 Physical and Mental Discipline: Forms of Thought and Practice

From the definition of *discipline*, it is apparent that the term constitutes a double-edged sword: (a) it specifies the organised ways in which scientists and practitioners go about their work such that they can indeed be identified in terms of specific practices; and (b) getting to the point of exhibiting these practices requires physical and mental discipline, generally instilled by imposing (more or less severe) constraints in the way people work. In fact in Foucault's genealogy, discipline arises historically through punishment, and emerges as a technology of power through surveillance that is increasingly internalised as self-surveillance. Arguably then, we are first disciplined by others before increasingly disciplining ourselves.

If we return to Fig. 3.2, but now consider the relation of the two alienated and separate activities of 'maintaining the discipline through teaching', and 'labour activity mediated by the discipline' then we see (in Fig. 3.3) some sources of new contradictions. Thus, the consciousness associated with learning actions (e.g. learning to read instruments) is now envisaged as providing instruments of productive labour (using instruments as tools to make consumables, say). However, there is plenty of room for things to go wrong here: typically the learning outcomes are not in fact fit for purpose, and the learners know it. Then an 'alienated' discipline is required to motivate the learning activity itself: institutions of learning become increasingly empowered to create and assert new motives for the learning of disciplinary knowledge.

Even military discipline, the epitome of discipline and its historical antecedent, was the result of increasing power through coordinated actions, involving physical, material, and behavioural standardisation of rigorous, detailed procedures—a model for schooling book-keepers and perhaps school mathematics. These disciplinary forms also achieve cohesion and esprit de corps, and thus implicate certain kinds of authoritarian disciplinary identities. The emergence of discipline in the military precedes but subsequently develops alongside schooling in the monastic tradition, and finally military schools emerged alongside formal, mass schooling, which constituted not only physical and mental discipline but also a system of social ordering (Foucault, 1978). For Foucault, discipline involves primarily a technology of control over the body, a 'microphysics' of power, which results in the docility of those so disciplined. Strict adherence to specified rules and linguistic forms subsequently constitute a self-imposed discipline.

The significant point here is that the practices involved in maintaining the discipline (its teaching and learning) become at least somewhat separated or even alienated from the activity systems where they are supposed to be practised productively. The disciplines take on a life of their own. This reached its ultimate form in mathematics, when pure mathematicians like Hardy declared the absolute lack of utility of their pure mathematics. And yet, there is hardly a branch of pure mathematics that might not turn out to be importantly useful, even though in some cases this may be some generations after the discipline has invented it. This has proved repeatedly the case now for some hundreds of years, from Boolean algebra and its eventual exploitation in digital technologies, to non-Euclidean geometries and modern Physics.

It has to be claimed, then, that the disciplining of mathematics that takes place by mathematical research communities is not reducible to an arbitrary discipline like that of the military. Its disciplinary criteria of certain habits of mind—elegance, proof and efficiency—appear to serve the discipline well in some functional sense, even though it can be argued that these have become independent of any immediate utilitarian, productive activity outside of mathematics itself. How has this happened, and how can one ensure that a discipline, e.g. a mathematical discipline, will continue in such a fashion? This question takes us deeper into philosophy than this chapter can go. But it is an important feature of the discipline that it continues to hold some of these properties and that this property may not apply equally well to other candidates for the title of discipline.

There is a strong case for questioning 'mathematics education' as such a functional discipline; many including ourselves have argued that our discipline has become so detached from mathematics proper as to be dysfunctional. This may be true of many other school subjects as well, and we should not reduce the value of interdisciplinarity of mathematics with other disciplines to that of the managerial integration of school subjects, even those that bear the same name!

3.6 Interdisciplinarity: Working Between and Across Disciplines

The problem of interdisciplinarity may be framed in terms of the activity theoretic approach outlined above. Here, paradigmatically, in contrast to the normal organisation in society, whereby products are exchanged by means of a generalized exchange form, i.e. money, two or more groups (organisations) representing different disciplines, may come together to work on a common object to result in a common product (Fig. 3.4). Thus, for example, one study reported how an interdisciplinary project emerged when three 'relatively autonomous project groups, composed of researchers with different disciplinary backgrounds' came together for the purposes of constructing 'the key parts of this projected production system: the development of microbial strains' (Miettinen, 1998, p. 430).

Here, readers should keep in mind that an activity is defined by the object/motive on which the collective is working, but the object is—like all the 'moments' represented as nodes in Figs. 3.2, 3.3 and 3.4—in the process of transformation: the work of the collective involves transforming the 'raw' objects into 'outcome' objects that meet a social need. In the classical case of labour activity, the actions of the various workers lead to the manufacture of finished consumables. Thus the 'motive' involved is the envisaged transformation of the 'raw' object into 'outcome', sometimes this motive is imagined, sometimes only emergent through the collective actions of the individual subjects involved. When we speak of 'object/motive' then, we have all this in mind: an activity is defined by the 'motive' of transforming an 'object' into a new form, an 'outcome' that meets a social need: that is, it has some kind of use value (which in the classic case of commodity production is the 'use value' of the commodity in the sense of Marx).

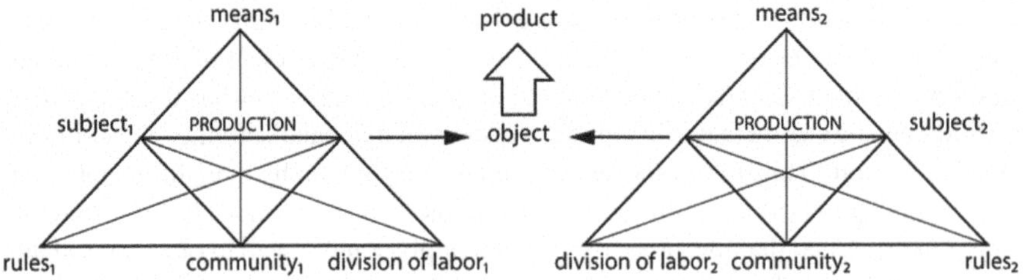

Fig. 3.4 In an interdisciplinary project, two different and otherwise autonomous activity systems collaborate with a *common* but temporary object/motive

Common object-motives often characterise interdisciplinary projects, even though the contributing disciplinary activity systems differ, each with its own distinctive characteristics, tools, and perspectives (Fig. 3.4); other forms of collaborations across disciplines may be organised differently, e.g. as in Fig. 3.3. The possible contradiction is immediately apparent. Because each part of an activity system is a function of the whole and is permeated by the characteristics of all other parts, the motives characterising any two activity systems may differ. That is, any interdisciplinary endeavour involves the work of specifying a *common* object/motive (product), which likely differs from object-motive1 and object-motive2 that characterise the respective mono-disciplinary practices (Fig. 3.4).

In the case that the two disciplines involved *are* able to come together in joint work on a common object, then it is possible to analyse their joint activity as a single activity system with a division of labour that incorporates both disciplines. In the simple case of joint work of a doctor and patient, the two may work seamlessly to prescribe and effect the treatment with a common object/motive of fixing the patient's medical problem; but the whole may fall apart into separate and contradictory activities if their motives are not well aligned, perhaps because the patient has other motives than their own health, they dislike the side effects of the treatment, or the drugs are too expensive, etc. In the education context the disjuncture may be even more strongly marked, as there is in general not much negotiation between the teacher and the learner as to the object of their joint actions: indeed it is disastrously common for the teacher to conceive of the object as *being* the learner. Another general problem in the education context is the misrecognition of the true object/motive. Thus, it is often stated that learning is the object of schooling, whereas in practice, the production of grades, grade reports, and diplomas is the actually realised motive toward which schools, teachers, students, and parents are oriented (Roth & McGinn, 1998).

A common case in point in education is the misalignment of the teacher who designed a classroom lesson and the teaching assistant who is assigned to help a group (usually a particular child or subgroup of the class): unless the teacher and assistant share understandings of the purpose of the lesson there may be essentially separate activities going on even in the mono-disciplinary classroom. (Even more sharp might be the case of the separation of the curriculum designer/author from those who teach, e.g. the teacher who uses a powerpoint designed by some other teacher with a different perspective!) Interdisciplinary teaching is even more prone to this, because the teachers of different disciplines have distinct disciplinary practices and learning outcomes in view.

The difficulty in defining a common object can often explain the failure of projects designed to be interdisciplinary. On the other hand, in successful projects, new object/motives may be created in such a way that they make sense within each of the disciplines (e.g. Miettinen, 1998). A good example of such an endeavour was observed in the collaboration of printers and designers to redesign the printers' workplace (Ehn & Kyng, 1991). Together, representatives from the two disciplines built mock-ups to model what happens in the workplace, and, in so doing, developed a new form of discourse that made sense within each discipline and constituted a sense-giving field that made sense across the fields.

One function of common objects (e.g. representational tools) is that they coordinate the activities involved even though the practices surrounding these objects differ. These objects are commonly known as *boundary objects*: such objects define boundaries between practices (forms of activities). Thus, for example, in the manufacture of an aircraft, many different disciplines are involved; the coordination between these very different disciplinary fields is achieved by means of drawings (Henderson, 1991). These drawings have different functions and are understood differently on the shop floor, in the accounting department, for the electrical engineers, or the inventory control department. Because of this, the object also may be thought of as a *conscription device*, that is, an entity that brings together (enrols) members of different disciplines (communities of practice) for the purpose of realising a common object/motive, which also is itself made visible by that same device.

There may also be (usually small numbers of) individuals who are familiar with and exhibit expertise in two disciplinary fields (Star, 1995). They cross and transcend boundaries, sometimes being called 'brokers', 'wizards' or 'gurus' that are highly competent in multiple domains and across multiple systems of formal representations. More prosaically, we are all multi-disciplinary in the loosest senses of 'discipline': we taxi-drive our children to school, become launderers on wash day, paint the garden fence on the weekend, nurse our old-aged when sick, etc. But these are riven with contradictions and power relations: some are paid, some confer esteem and cultural capital, and all are shot through with social divisions along the gender, race and class divides.

3.7 Interdisciplinary Power and Conflict

Anyone working in academia knows about the institutional hierarchies between and within faculties and forms of knowledge. Thus, the natural ('hard') sciences tend to be regarded as higher in esteem and more powerful than the social ('soft') sciences (Bourdieu, 2000); within a particular field, the same gradations are reproduced, e.g. in psychology, there are gradations from the 'hard' (e.g. experimental psychology in the lab) to 'soft' (counselling psychology); and within each field there are gradations, where some scholars are on top of the heap and others are the new underclass of 'proletarianised intellectual' on zero hours contracts. The disciplinary divisions between hard and soft sciences found a parallel in gender divisions, which was the result of institutional practices that systematically excluded women from the natural sciences, especially hard, natural sciences (Shumway & Messer-Davidow, 1991). Similar power differences have been recognised within multi-disciplinary teams in inter-professional contexts (e.g. between medical doctors, nurses, and assistants for instance).

One of the conditions for interdisciplinarity to function effectively is the active management of the historically developed attitudes between disciplines and forms of inquiry for the purpose of overcoming condescending and colonising attitudes (often amplified by racism, sexism, class etc.). The structures of power gradations that separate the faculties and disciplines—while they have a degree of autonomy—are homologous with the entire field of power in society at large: natural sciences (STEM) being opposed to faculties of social sciences (Bourdieu, 1984). The ruling relations within disciplines reproduce those between faculties. Knowledge constitutes a form of 'symbolic capital' that may be accumulated as any other form of capital. In the sociology of symbolic capital, the university faculties are characterised by their position within the historically evolved academic field of power, each with its own internal field of power and cultural capital. Within disciplines, certain schools, sometimes associated with specific universities (e.g. Ivy League or Oxbridge) reproduce these structures through patronage and graduate student exchange (e.g. Traweek, 1988).

The practices of selection and indoctrination *within* each discipline contribute to the reproduction of differentiation *between* the disciplines. Cultural capital contributes to the constitution of a discipline within society as a whole and to the relative status of the individual within the discipline. This entire disciplinary formation therefore acts as a great 'weight of the world', making for difficulty in expecting those 'schooled' and 'disciplined' in one field to relate in effective ways with others whose habits have been formed in relatively independent, and contradictory fields. Thus alienation of the discipline from productive labour provides the conditions for strengthening the alienation between disciplines, producing the conditions for ever greater barriers to interdisciplinary work in the academy/school.

3.8 Transdisciplinarity: Considerations of Dialogism, Heteroglossia, and Voice

In everyday life and in most productive work activity, one is rarely conscious of the multiple disciplines that have played a part historically in the concepts, tools and objects that we engage with. Getting on with the task, we ignore the mathematical work that went into the design and manufacture of the tools we use (the computer), the rules of the systems we unconsciously obey (the timetable), even the mathematical concepts embedded in the discourses that mediate our work (there isn't enough time to list them all). In getting on with life, then, we transcend the disciplines, and our activity is transdisciplinary in the sense that we are not conscious of the disciplinary moments of the activity that are hidden in black boxes. But if we look closely and reflect, we can begin to see the many disciplines mediating our work and discourse, and a heteroglossia of such disciplines in everyday transdisciplinarity. The idea of heteroglossia is well exemplified in the preceding example of the printers and designers, who began the interdisciplinary project by developing a hybrid discourse in which *that* workplace redesign was accomplished (functioning much like *Sapir*, the hybrid language of Mediterranean merchants permitting trade across nations, cultures, and mono-languages). Otherwise, the two activity systems remained distinct and maintained their disciplinary discourses.

To add to the previous account, then, one should consider Bakhtin's conceptualisations of voice, dialogism, and heteroglossia: many in activity theory consider these Bakhtinian concepts to be part of the modern version of cultural- historical activity theory. Given Bakhtin's view of the social world, we can consider a 'discipline' as a language genre—or, equivalently, a Wittgensteinian *language-game*, defined as a language and the practical activity with which it is interwoven—in which certain structures of utterance are formed with particular functions and audiences in mind: there is a natural assumption that persons in modern cultures speak with many voices, adopting many such genres as appropriate, within a polyphony of discourses. In this view we can all perhaps be expected to be multivoiced, or multidisciplinary in a discursive sense.

We can now draw on Bakhtin's conceptual tools: in Bakhtin's framework an author's power is 'monologic' when it demands that the addressee adopts the voice and genre of the speaker, and when the addressee is thereby ventriloquated by the authority (Bakhtin, 1981). On the other hand in ethical, 'free', but effective communication, there should be a dialogism between speaker and addressed, in which the speaker seeks to offer the addressee language that can become 'internally persuasive'. In this view, effective multidisciplinary communication would require speakers to at least partially adopt voices in the language/discipline of the 'other' for others who 'speak' a different discipline's discourse, or for instance, a different genre of mathematics (Williams & Wake, 2007). In Wenger's (1998) conception of 'knowledgeability' within a landscape of communities of practice, to be effective does not require one to be an old-timer or competent practitioner of adjacent boundary communities but one needs to be just knowledgeable enough of the practice to be effectively competent with one's own. In the school context, this means that the teacher needs to understand the classroom to some degree from the perspective of the teaching assistant, and vice versa, even though these two professionals may be considered to belong to different professional communities/disciplines and prefer different language genres.

Together with Bakhtin's concepts, this more positive and hopeful conceptual framework might have affordances that will support understandings of multidisciplinarity using ideas of the broader 'social language' in which distinct language genres are formulated, and make use of his notion of dialogism to think about talk across disciplines (where one's word is always half the other's, i.e. 'someone else's') and even 'hybridity' of disciplines. A hybrid utterance according to Bakthin contains at least two interanimating voices in ways that allow new meanings—this can occur in dialogues in ways that produce creativity, and novelty—typified in his analyses of humour. This exemplifies the hybrid utterance, because the hybridity occurs at the level of meaning of the utterance; but still one can discern the two separate voices and language genres within the utterance. This does not involve, arguably, a new 'hybrid language or language genre' such as that which might occur when two disciplines create a new one, such as, perhaps 'biochemistry', or even 'engineering-mathematics'. We suggest however that what occurs when specialist mathematicians or scientists collaborate across disciplinary boundaries in some task might be usefully examined from this perspective. The key point is that each discipline does not need to incorporate the expertise of the 'other', but to have just enough meta-disciplinary knowledge about the nature of its own discipline in its relation to the other disciplines and to the task to engage in an effective dialogue (i.e. one where the object can be adequately formulated by both disciplines). This is just what is required to speak in hybrid utterances that include the other disciplines, at least where the object of activity is concerned, without necessarily forming a new hybrid discipline.

In the account of interdisciplinarity given above, we call this metadisciplinary knowledge, and the suggestion in the Bakhtinian frame is that this might best be studied (and developed) in a transdisciplinary context of activity where each discipline rubs up against and is subsumed within a wider activity, possibly with other disciplines also. What determines whether the two disciplines are engaged in the *same* activity, ultimately, is that they share a common object-motive, even when they maintain their own disciplinary tools, within a division of labour that includes both disciplines. This requires enough meta-knowledge to negotiate their motives with 'others', which includes understanding what each discipline can offer.

3.9 Identities in Disciplinary and Interdisciplinary Practices

The focus on identity in much social theory might similarly be significant in many inquiries, particularly those that are interested in learner engagement, choice-making in further and higher education, and career choices (i.e. learner identities) but also professional identities (e.g. teachers, educators, health professionals). The consequences of learning to see oneself through reflective activity as a 'certain kind of person' (perhaps a scientist or mathematician of a certain kind) are thought to be crucial here: but these reflections are always reflections 'on' past experiences, and particularly how 'others' have positioned oneself in joint activities. Thus we have been told by professors of mathematics in prestigious universities that 'most of them (i.e. their undergraduate mathematics major students) won't become mathematicians', positioning their graduating students as non-mathematicians even though they are likely to be employed as future teachers of mathematics, users of quantitative methods in industry, etc. In this regard then, we consider the need for a social theory of identity-making in practices of various kinds, including in dialogues with powerful others who position subjects in activity, as well as reflective discourses with oneself. Clearly disciplinarity and interdisciplinarity mediate these: and whether an engineer or a scientist sees themselves as some kind of mathematician might be a key question to ask when mathematics is but one discipline in an interdisciplinary or transdisciplinary endeavour. Compared with professional identities, and post-compulsory students' learner identities, school students tend to have relatively weak attachments to the disciplines. It is then more important to many that the disciplines be shown to be of some use, than that they become inducted in the practices of the discipline per se.

Consider the example of an interdisciplinary mathematics, science and technology project in a secondary school engaging the whole cohort of 150 students aged 11–12 years. In an attempt to motivate the students with an 'outside academia' task, the students were asked to develop a plan to improve their school environment: making their break times in school more pleasant. It was agreed that the winning project would be funded, imposing a budget of £500, to make sure the project would actually be implemented; and there was to be a vote of the whole cohort to decide on the winning project. In order to justify the timetable given over to the joint project, each subject-department led some teaching of their discipline (actually their curriculum) in ways intended to help the children consider relevant plans. There were lessons about: the science of plants and kinds of soil needed for planting; how to handle data from a survey to find out what other children would like from such a project; and regarding technology, how to design and make new furniture. Thus, each curriculum area ticked a box that the project met some of their curriculum objectives for the year, even though a great deal of the time was not spent focused on the substance of the curriculum as such.

However, of all the projects considered, the winner in the end was one that used little disciplinary knowledge, and apparently none of the science taught—much in the same way as a case described by Leontiev, which was the outcome of an inappropriately framed and chosen object/motive. The winning project involved the building of a path that would allow students to access the school directly from the public busstop, thus saving them a ten minute walk round the streets to the front entrance of the school. Transdisciplinarity appeared to win here: the students learnt important meta-disciplinary knowledge—that there may be no place for science in practice, and the benefits of an un-disciplined approach that is free to focus on the actual problem in hand. In this light, this lack of discipline can be a good thing, maybe even an essential requirement, as long as it is well informed. One has to have some knowledge about the discipline (precisely meta-disciplinary knowledge or know-how) in order to know when it may have little or no place in informing an issue. A pity there is no room in most school mathematics curricula for this.

The interdisciplinary categorisations described at the beginning of the paper suggests that transdisciplinary problem solving can, on reflection, produce meta-disciplinary knowledge that negates disciplinary power (to some extent) and introduces the expansive cycle in Fig. 3.5, where the second cycle involving the mono-inter-trans-disciplinary knowing becomes a self-aware knowing, in light of the meta-knowledge. The kind of knowledge involved in this cycle certainly includes the understanding of mathematics as 'mathematical modelling' for instance. But it also would include understandings of the particular historical relationship between mathematics and physical sciences or engineering; and in the case of the profession of nursing, it might involve the situated appreciation of the relationship between risk and probability, and so on.

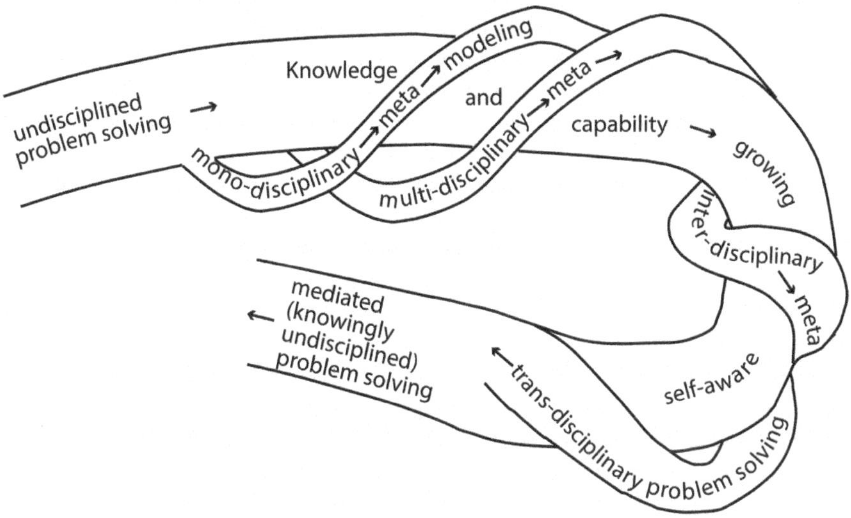

Fig. 3.5 An expansive meta-cycle of interdisciplinary problem solving

3.10 Conclusion: Social Theory for Interdisciplinarity

Activity theory (after Vygotsky/Leontiev) has given us some powerful tools to think about interdisciplinarity and the relations (contradictions) between learning disciplines and productive labour. The theory does not emphasise the structural relations with capital and power (implicating alienation) offered by the Bourdieu/Foucault perspectives, and which make salient the powerful alienation of disciplines from superseding activity, insofar as they emphasise (a) the arbitrary quality of capital (Bourdieu) and (b) governmentality of disciplines (Foucault). They also raise the possibility and shape of certain forms of resistance to capital (reflexive sociology) and disciplinary discourses (becoming aware) respectively. These two perspectives in combination might offer insights that might reveal some of the potential for interdisciplinary work but also may help explain its widely recognised failures. The concepts from Bakhtin helped to envisage a transdisciplinarity in which multi-disciplines work together 'dialogically', but in which the disciplines might rise to consciousness through reflective activity: hence the emergence of metadisciplinarity through transdisciplinarity. It also gives us some additional insights into activity through concepts such as language genre, mono/dialogism, and internally persuasive discourse.

Being undisciplined in the pre-historic state is a matter of absence of discipline. But proceeding historically to a disciplined culture, we can now negate this in a disciplined consciousness which is nevertheless to a degree free to be undisciplined, i.e. to one where we are aware of the disciplines (maybe even specialised and skilled in some of them) but thereby relatively un-disciplined by them. This is a new perspective on being 'undisciplined', i.e. being aware of making us somewhat free of disciplines.

Rather than making interdisciplinarity the new scientific dogma or ideal practice to be achieved, a more productive approach may consist in situating inquiries and endeavours according to the complexity of the questions asked. Thus, on the scale of complexity, interdisciplinarity may actually be thought of as a continuum of relations between disciplines, between mono-disciplinarity, on the one end, and trans-disciplinarity, on the other end, with multi/inter-disciplinarity and trans-disciplinarity offering more or less hybridity of the disciplines involved (Collen, 2002). As a result, neither mono-disciplinarity nor any other 'level' is ever eradicated or even invalidated; indeed the core value of the discipline may provide precisely the value to other disciplines that interdisciplinarity requires. But the continuum allows inquirers, who have the awareness, to advance by moving towards more complex inquiries involving more than one discipline in ways that lead to advances and novel forms of insights (e.g. Hicks, 1992) or to return to less complex inquiries to draw on the advantages that arise from lower levels of complexity (in objects, organisational forms, efforts).

The issue of meta-disciplinarity has hardly been explored in educational research and we suggest that this deserves a great deal more attention in the context of problem solving in general and multi-disciplinarity in particular. In this, inter-disciplinary work and education parallels the work in inter-professional work and education in contexts such as the health services, and this highlights the relation of the multi-disciplinary team to expand to include the user.

Additionally we have turned particularly to Bakhtin as a perspective on language genres to think about 'disciplines' as characterised by their disciplinary 'languages' or 'genres' and voices, but also to think of disciplines as if they actually were discourses (language-games). Then an interdisciplinary project or activity becomes metaphorically conceptualised as a hybrid utterance containing two interanimating voices, with new hybrid disciplines emerging as hybrid genres or even languages like Sapir or Creole. We suggest that this conceptualisation might be productive for future work in the field. We do argue however that the main concerns that the academy needs to address are to be found in the alienation of disciplinary activity (and thus subjectivity, identity, etc.) not only from other disciplines but also from productive labour, and specifically from 'real' inquiry and problem-solving in learners' interests. We concluded that being 'disciplined' should be challenged by the needs of labour, becoming 'undisciplined' but in a new, disciplined way, superseding disciplinary and metadisciplinary knowledge.

3.11 Coda

In a review, Williams et al. (2016) concluded that interdisciplinary mathematics education in the context of inquiry and problem solving offers mathematics to the wider world in the form of added value (e.g. in problem solving), but on the other hand also offers to mathematics the added value of the wider world. In this chapter we build on this but take the argument further and begin to emphasise the need for learners to become educated in the nature of the discipline of mathematics and of its relation with other disciplines, freeing the problem solver to some degree to be knowledgably un-disciplined, an essential requirement of good problem solving. In this move, theoretical resources included (a) activity theory, (b) language genres and Bakhtin, and (c) Bourdieu and Foucault.

The first involves the separation of disciplinary activities of mathematics (e.g. learning, teaching, researching the discipline itself) from their use in an expansive system of productive problem solving in practice (when the production system and other disciplines may be critical). The second involves hybridity, identity, and disciplinary discourses as language genres. The latter takes us to (field and discursive) theories of power and resistance. In all, these theoretical perspectives help us to understand how to research interdisciplinary mathematics and its cultivation in academe. But, just as we do not argue that all academic activity must be interdisciplinary, so also we do not suggest that all this theory is required for every particular task in educational research and practice. Instead, we show how these theories offer researchers and practitioners some theoretical tools, some of which, or combinations of which, can be fit for the purposes of understanding interdisciplinary activity. For instance, future research and practice will ultimately determine whether, or to what extent, and how activity can be both disciplined and undisciplined, i.e. knowingly so. We argue and conclude that one key to this knowing in the interdisciplinary context is in the expansive, meta-cycle above.

References

Bakhtin, M. M. (1981). *The dialogic imagination: Four essays*. Austin, TX: University of Texas Press.

Bourdieu, P. (1984). *Homo academicus*. Paris, France: Les Éditions de Minuit.

Bourdieu, P. (2000). *Pascalian meditations*. Stanford, CA: Stanford University Press.

Collen, A. (2002). Disciplinarity in the pursuit of knowledge. In G. Minati & E. Pessa (Eds.), *Emergence in complex, cognitive, social, and biological systems* (pp. 285–296). New York, NY: Springer.

d'Ambrosio, U. (1990). Literacy, matheracy, and technoracy: A trivium for today. *Mathematical Thinking and Learning, 1*, 131–153.

Durkheim, É. (1893). *De la division du travail social* [*Division of social labour*]. Paris: Felix Alcan.

Edwards, A. (Ed.). (2017). *Working relationally in and across practices: A cultural-historical approach to collaboration*. Cambridge, UK: Cambridge University Press.

Ehn, P., & Kyng, M. (1991). Cardboard computers: Mocking-it-up or hands-on the future. In J. Greenbaum & M. Kyng (Eds.), *Design at work: Cooperative design of computer systems* (pp. 169–195). Hillsdale, NJ: Lawrence Erlbaum Associates.

Engeström, Y. (1987). *Learning by expanding: An activity-theoretical approach to developmental research*. Helsinki: Orienta-Konsultit. https://doi.org/10.1017/CBO9781139814744.

Foucault, M. (1970). *The order of things*. New York: Random House.

Foucault, M. (1972). *Archeology of knowledge and the discourse on language*. New York: Pantheon Books.

Foucault, M. (1978). *Discipline and punish: The birth of the prison*. New York: Vintage Books.

Foucault, M. (1988). *Madness and civilization: A history of insanity in the age of reason*. New York: Vintage Books.

Henderson, K. (1991). Flexible sketches and inflexible data bases: Visual communication, conscription devices, and boundary objects in design engineering. *Science, Technology and Human Values, 16,* 448–473.

Hicks, D. (1992). Instrumentation, interdisciplinary knowledge, and research performance in spin glass and superfluid helium three. *Science, Technology and Human Values, 17,* 180–204.

Leontiev, A. N. (1978). *Activity, consciousness and personality*. Englewood Cliffs, NJ: Prentice Hall.

Leontiev, A. N. (1981). *Problems of the development of mind*. Moscow: Progress.

Lynch, M. (1985). Discipline and the material form of images. *Social Studies of Science, 15,* 37–66.

Marx, K., & Engels, F. (1974). *The German ideology*. London: Lawrence.

Miettinen, R. (1998). Object construction and networks in research work: The case of research on cellulose-degrading enzymes. *Social Studies of Science, 28,* 423–463.

Roth, W.-M., & McGinn, M. K. (1998). >unDELETE science education:/lives/work/voices. *Journal of Research in Science Teaching, 35,* 399–421.

Shumway, D. R., & Messer-Davidow, E. (1991). Disciplinarity: An introduction. *Poetics Today, 12,* 201–225.

Star, S. L. (1995). The politics of formal representations: Wizards, gurus, and organizational complexity. In S. L. Star (Ed.), *Ecologies of knowledge: Work and politics in science and technology* (pp. 88–118). Albany: State University of New York Press.

Traweek, S. (1988). *Beamtimes and lifetimes: The world of high energy physicists*. Cambridge, MA: MIT Press.

Wenger, E. (1998). *Communities of practice: Learning, meaning, and identity*. Cambridge: Cambridge University Press.

Williams, J., Roth, W.-M., Swanson, D., Doig, B., Groves, S., Omuvwie, M., et al. (2016). *Interdisciplinary mathematics education: State of the art*. Cham: Springer. https://doi.org/10.1007/978-3-319-42267-1.

Williams, J. S. (2016). *Becoming un-disciplined with science and mathematics*. Downloaded July 2018 from http://stemedcon.deakin.edu.au/wp-content/uploads/sites/25/2017/05/Proceedings-Julian-Williams.pdf.

Williams, J. S., & Wake, G. D. (2007). Metaphors and models in translation between college and workplace mathematics. *Educational Studies in Mathematics, 64*(3), 345–371. https://doi.org/10.1007/s10649-006-9040-6.

Williams, J. S., Wake, G. D., & Boreham, N. C. (2001). College mathematics and workplace practice: An activity theory perspective. *Research in Mathematics Education, 3,* 69–84.

Chapter 4
Integration from a Commognitive Perspective: An Experience with Mathematics and Music Students

M. Alicia Venegas-Thayer

Abstract Since one of the keywords in the interdisciplinary discourse is integration, the aim of the study was to describe the actions of participants that could be considered as part of an integration process involving mathematical and musical discourses. Based on the commognitive perspective developed by Anna Sfard, which argues that communication is a collectively performed patterned activity, here integration is a way to develop a new type of communication. Music and mathematics students participated in an experience, where it was possible to observe how line graphics of random data were interpreted through actions from a musical discourse and how the students developed a new form of communication when talking about chords, which they called "baggies".

Keywords Commognition · Interdisciplinarity · Innovation · Higher education · Random variable

4.1 Introduction

During the 1950s, architect and composer Iannis Xenakis (1922–2001), developed a method for musical composition called 'Stochastic Music'. It is music constructed from the principles of indeterminism, where the laws of probability become a necessity in the process of composition (Xenakis, 1963/1992). His music works are characterized as sound clouds, where every sound particle contributes to the work as a whole (Abornés & Milrud, 2011). His motivation was not the need to include mathematics in the process of musical creation, but rather to consider music and sound as media and thus allowing the materialisation of creations of the human thought (Xenakis, 1963/1992). Using mathematical notions of probability, algebra and group theory, Xenakis defines the transformations that each sound entity will have throughout the playing. This music-mathematical relationship is widely described in his book called 'Formalized Music' (Xenakis, 1963/1992).

M. A. Venegas-Thayer (✉)
Pontificia Universidad Católica de Valparaíso, Valparaíso, Chile
e-mail: aliciavenegasth@gmail.com

B. Doig et al. (eds.), *Interdisciplinary Mathematics Education*,
ICME-13 Monographs, https://doi.org/10.1007/978-3-030-11066-6_4

The use of technology has led composers to create works that in other ways would not be possible, where mathematical and physical models together with these tools are now part of the musician's problematic (Cádiz, 2012; Ferreira, 2001). This change in practice and musical discourse opens the door to new forms of composition and musical production in general, where mathematicians and musicians can carry out a creation process interacting collaboratively. Examples of contemporary composers are James Tenney, developer of an algorithm for the composition called dissonant counterpoint, considering a statistical and probabilistic background (Polansky, Barnett, & Winter, 2011). An other example is Michael Winter, whose processes of composition and studies in musical theory are characterised by the collaboration not only with other artists but also with mathematicians and scientists. This musician-mathematical relationship is also reflected in the analysis of works and musical performances in which advanced mathematical procedures have been used for scientific music research in music theory (Noll & Peck, 2007).

In these new scenarios, musicians and mathematicians have been introduced to practices and discourses that are not necessarily their specialty. As Boix (2010) describes it, participants of interdisciplinary processes have to "*integrate* information, data, techniques, tools, perspectives, concepts, and/or theories from two or more disciplines to craft products, explain phenomena, or solve problems, in ways that would have been unlikely through single-disciplinary means" (p. 289). During the process, the participants engage in a collaborative dialogue, "including debates and conflict, which both transforms the understanding of individual participants and produces new knowledge, new solutions, and even new disciplines that would not be possible without such dialogue" (Derry & Schunn, 2014, p. xiii). In other words, those who make these new productions possible are the musicians and mathematicians involved in the process, whether through cross-dialogue with other specialists or through personal processes of study and research.

Since one of the keywords in any interdisciplinary discourse is *integration* (Klein, 2010; Newell, 2001), the aim of the study reported here was to describe the actions of participants that could be considered as part of an integration process involving mathematical and musical discourses. The framework of this research is the Commognitive Theory (Sfard, 2008, 2012), adopting a communicational perspective about thinking. According to Sfard (2008), communication—and therefore, thinking—is possible because of the habit of reacting to certain actions with certain kinds of reactions, so here *integration* is defined as a way to develop a new type of communication. Music and mathematics students participated in an experience, where it was possible to observe how line graphics of random data were interpreted through actions from a musical discourse and how the students developed a new form of communication when talking about chords, which they called "baggies".

4.2 Commognition: Thinking as Communication

Starting from the premise that humans are socially committed from the day they are born, Sfard (2008) points out that human development is based on communication. As she describes it, thanks to communication, human beings have been able to satisfy from the basic necessities that keep them alive, to the most advanced and complex cultural demands. Built on the Vygotskian socio-constructive perspective where the "historically established, collectively implemented activities are developmentally prior to all our uniquely human skills" (Sfard, 2008, p. xvii), the author concluded: if human thinking is one of these skills, the most obvious collective predecessor is interpersonal communication (Sfard, 2008). She proposes a new way of thinking about thinking, pointing out "thinking can be usefully defined as the activity of communicating with one self" (Sfard, 2012, p. 2). The author defines communication as,

> a collectively performed patterned activity in which action A of an individual is followed by action B of another individual so that (1) A belongs to a certain well-defined repertoire of actions known as communicational, and (2) action B belongs to a repertoire of re-actions that fit A, that is, actions recurrently observed in conjunction with A. (Sfard, 2008, p. 296)

That is, "what makes communication possible is the fact that the community got into a habit of reacting to certain actions with certain types of re-actions" (p. 88). Then she makes clear that this action and reaction relation does not only depend on the A action, but other factors, like the historical moment where A and B are performed, that is the context, the previous actions, the identities of the participants, and so on. Thereby, cognitive processes and communicative processes are actually "different (intrapersonal and interpersonal) manifestations of the same phenomenon" (p. 296), and Sfard integrates these under the term *commognition*.

From this definition, Sfard (2008) deduces the existence of different *types of communication* or types of commognition, called *discourses*. Each discourse is characterised by: *keywords, visual mediators* - those visible objects used in any process of commognition and symbolic artefacts from specialised discourses like science or mathematics - *distinctive routines* or patterns of how to perform some tasks, and their *endorsed narratives*, those that the community has labelled as true (Sfard, 2008, 2012). There are no explicit boundaries between one discourse and another and there is even a mutual collaboration. Nevertheless, they may become sufficiently distinguishable from each other. The discourses are not stable entities, but on the contrary, they change over time. Changes in a discourse mean changes in the way of communicating, which directly influence human practices, and vice versa, changes in practice imply changes in discourse (Sfard, 2008, 2012).

According to these definitions, Sfard (2008) argues that *mathematical thinking*—or *mathematics*—is a discourse, with the following characteristic features: *keywords* (three, triangle, set or function, among others), *visual mediators* (numerals, algebraic symbols, and graphs), *distinctive routines*—patterned ways in which mathematical tasks have being performed—and generally *endorsed narratives*, that is theorems, definitions and computational rules (Sfard, 2008, 2012). In the same way, one can distinguish the discourse of algebra, of geometry, and of functions (Sfard, 2012). On the other hand, there are discourses about music and sounding phenomenon, with their own *keywords* (pitch, chords, musical scales or harmony), *patterned ways* in which musical task are performed, *visual mediators* (music stave, notes values) and it is also possible to talk about *endorsed narratives* accepted by a musical community, such as principles about harmony, just intonation or Pythagorean tuning.

Those "individuals capable of participating in a given discourse" (Sfard, 2008, p. 299) constitute the *community of discourse*. Sfard distinguishes between a person who belongs to this community, an *insider*, and a "person incapable of participating in the discourse" (p. 300), an *outsider*. The distinction and consideration of both perspectives can enrich the analysis of the discourse in question. According to Sfard, "what is senseless or inexplicable in the insider's eyes may become meaningful for an outsider, if only because from the outsider's perspective, the rules of the discourse in question do have alternatives" (p. 279).

As was mentioned above, *integration* is one of the keywords of any discourse about interdisciplinarity. So, this research has focused on characterising integration from the actions of individuals participating in an interdisciplinary process. To do so, an operational definition is required. Drawing from commognitive discourse, *integration* is defined as a process through which one or more individuals develops a new discourse by adapting and modifying keywords, mediators and/or procedures belonging to different discourses in a way that allows them to achieve a specific goal.

4.3 The Interdisciplinary Collaboration Experiences

If the discourse is a type of communication, this is characterised by a set of actions and reactions practised by the discourse community, so "development of discourse is, by definition, a product of collective human actions" (Sfard, 2012, p. 02). To describe those collective human actions associated to an integration process, an experience for university students of music and mathematics was organised, inspired by the methods of Stochastic Music composition developed by Iannis Xenakis. The problem proposed to the participants was as follows: *compose a piece of music from the data obtained by repeatedly throwing a fair coin.* It was raised in order to promote dialogue among the participants, so that they have to decide how to proceed and thus respond to the challenge. It was designed to be a two-hour session, beginning with a brief description of stochastic music, where they heard part of the piece "Metastasis" from Xenakis (1954/1965, track 3). Then, the problem was presented without giving further explanation, inviting participants to empower themselves through the problem by addressing it in their own way. Since it was planned for a session, it was expected that the participants would define a strategy for a musical composition, rather than compose a piece of music.

There were no preconditions about music and mathematics knowledge stipulated to participate. Two mathematics students and three music students accepted an invitation to participate in a two-hour activity, independent from their curricular requirement and schedule. The five students belonged to the same university in Valparaíso, Chile. Two experiments were organized, five days apart. For the first one, the author of this report participated as a practitioner in mathematics, together with a music student, who specialised in musical composition. For the second, two music students and two mathematics students, currently enrolled in their third or fourth year of their careers, were involved.

One situation from each experience was selected for this chapter, where it was considered that the participants' actions were an integration process. In both cases, the participants interpreted a characteristic feature of one discourse in terms of the other. The first one shows how the musician described a line graph in terms of musical discourses, and the second shows the students using the keyword "baggie" when talking about chords.

4.3.1 First Experimentation of Interdisciplinary Collaboration

Before the session, the mathematician prepared a spreadsheet of a simulation of one thousand tosses of a fair coin with results coded in terms of 0 (head) and 1 (tail). She organized the spreadsheet in tables and graphs, where it was registered the results of each toss, the absolute and relative frequencies and the difference between the number of heads, and tails were included.

Trying to provoke some interest in the data, the mathematician shows to the musician the graph of the differences between the number of heads and tails (Fig. 4.1). The musician points out that the graph could be a little melody. While indicating to the highest vertex of the graph, he says that this could represent the climax of the melody. He explains that any melody, no matter how complex it may be, is a combination of moments of tension and release. To exemplify this, he improvises a solfeggio while assigning notes to the vertex of the previous graph, starting with the notes C, E, and G. The highest notes, such as G, are the tension, after that, they go down and a release begins. The Fig. 4.1 represents what the musician described from the graph of the differences, but this is not an illustration built during the session.

During this conversation, he also suggests that the graph can also be *inverted* and says: "I think, this does not alter the mathematical relationship, because you are seeing it as in a mirror. It's like you have a mirror here and you're looking at it on the other way".

After almost an hour of conversation, the mathematician (MA) proposes to systematise the ideas they have been working on. The musician (MU) asks her to share the spreadsheet and the following dialog takes place:

MA: From this spreadsheet, what information do you prefer?

MU: Graphs. Ideally, graphs, you know? Because… yes, I'm going to see the numbers… I'm going to do it! But, it is going to take me some time.
MA: No. it's OK.
MU: But this is visual. And especially because I can see the distances, you know? For example, the distance from here to here [pointing to the tosses number 8 and 12 from the Fig. 4.1]

(…)

MU: Where I see more variation between tension and release, I would assign a melodic value. The music is very graphic. If I show you sheet music, you will see that the tension parts are below or above, but they can be differentiated. That is a climax, is very clear (Fig. 4.1), little weird because it usually goes lower, but still good.

(…)

MU: For example, the graph below is also good… the one that you have below (Fig. 4.2)
MA: That one has more coin tosses.
MU: Leave it… ¡Look! This is very good too. It is a classic little movement. We should have to assign a time value.
MA: OK.
MU: And that time value, we could also assign it with another graph, you know?

The musician uses the expression "melodic value", when the graphic can be interpreted in terms of pitch, meaning that the graph can represent a sequence of pitches, like in Fig. 4.1. The "time value" expression, he related to the duration of each note, but this kind of value was not exemplified on any graph in particular and it was just mentioned.

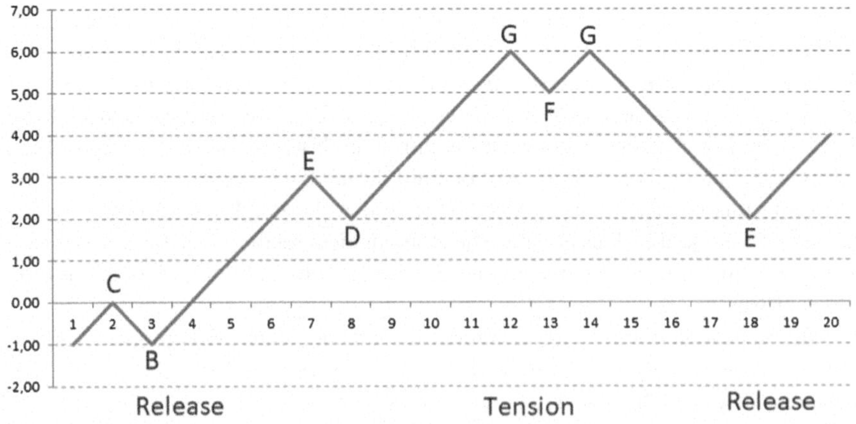

Fig. 4.1 Representation of the musician's example about tension and release

The participants decided to have a second session, which was not planned at the beginning of the experience. In that session, they focused on exploring and selecting a line graph that they found *melodically interesting*. Once they selected the graph, they create a spreadsheet to convert these random geometric shapes into pitches (A, A# B, C, C#, D…) and notes values (semibreve, minim, crotchet and so on). Finally, the participants transcribe these results to a staff using Version 2011.r2 of Finale (2011) and program them in Csound (Version 5.0; Vercoe & Fiftch, 2005), an audio programming language.

Figure 4.3 shows the line graph selected by the participants and the musical score obtained from that graph. To corroborate his idea about music being "very graphic", the musician makes a comparison between the curve made by following the heads of the musical notes on the sheet and the form of the line graph, showing that both have very similar shapes.

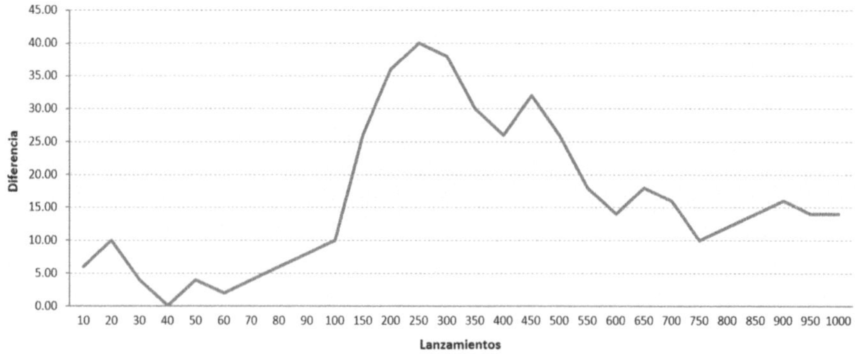

Fig. 4.2 The difference between the number of heads and tails over 1000 results

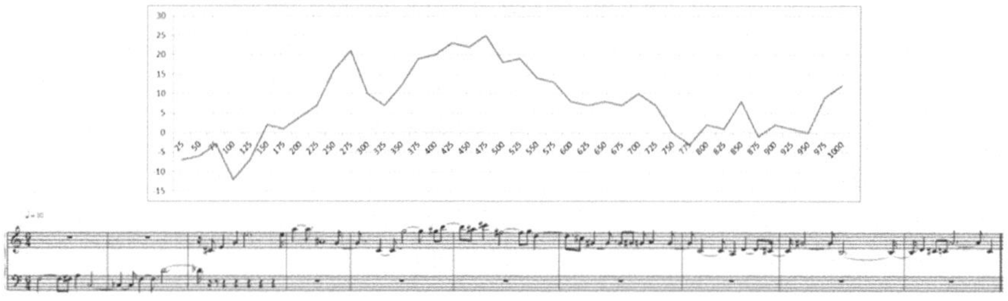

Fig. 4.3 The students' selected line graph and its interpretation as a melodic piece on sheet music

4.3.2 *Second Experimentation of Interdisciplinary Collaboration*

Two music students, MU01 and MU02, and two mathematics students, MA01 and MA02, participated in the second experimentation. After the participants hear the Xenakis music piece, MU01 makes a description of the composition process, stating that Xenakis must have made a list with all the options for each aspect of music—timbre, pitch, intensity, dynamics and duration. This utterance makes sense to MA01, because this kind of action is usually part of any activity in probabilities, that is, to recognise all the possibility when studying a random experiment. After 10 min of conversation, where the musicians showed the mathematicians a variety of factors to consider when composing, the students decided to make a composition for three kind of instruments: melodic, rhythmic and harmonic. To write the score and to simulate the instruments during the process, they used a scorewriter program called Sibelius (2011).

During the session, the student who led the group is the musician MU01. At the beginning of the composition process, he makes a list of the note values, its corresponding rests (intervals of silence) and a list with the 12 notes of the tempered scale (Fig. 4.4). The students will use these lists during the whole activity, which will allow them to identify the options available to define rhythms and melodies for the composition. They also agree that for each instrument, they are going to calculate only 10 notes or chords, just to check whether the procedure is right or not.

The following description focus only on the piano, particularly the creation of chords from random data. Since the piano is the harmonic instrument of the composition, they have to define chords instead of sequences of notes, that is, sets of two or more notes played simultaneously. At the beginning of this task, the participants present a series of ideas to do this. For example, MU02 suggests considering three-note chords; MU01 proposes focusing on the fingers of the hand and their position in the piano instead of the musical notes; MA02, to do a combination of 4 notes from 12 to determine all the four-note chords possibilities. Then, MU01 explains the structure of the piano as sequences of 12 notes repeated in different octaves; MA01

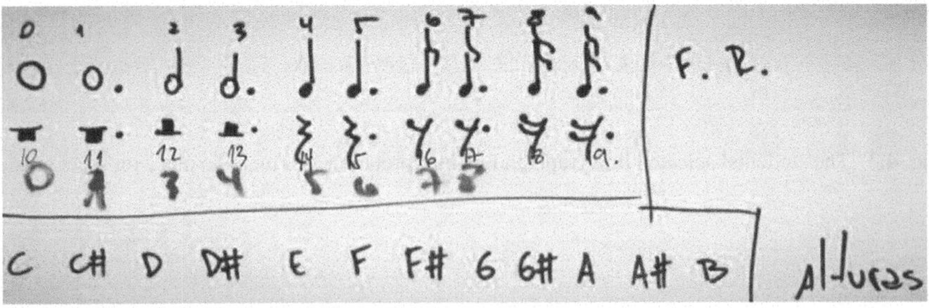

Fig. 4.4 List made by MU01 with the note values, rest values and pitches used for the composition process

suggests to determinate the notes for each hand, because a hand can reach the twelve notes of an octave. MU01 proposes to use four-note chords, while he is drawing four vertical marks attached to a kind of ellipse (first ellipse on Fig. 4.5) and MU02, to consider all those notes with the same duration.

During this conversation, MA01 asks about how many kinds of chords exist, and the musicians explain there are no restrictions on the pitches of a chord and the only restriction is the number of notes, because this cannot be more than the number of fingers from both hands. Then the interchange continues mainly between MA02 and MU01 about the number of notes that a chord could have. MA01 intervenes as shown in the following dialogue:

MU01: We have the possibility that… it will be… look it is going to be ten chords.
MA01: We must do… we must do…
MU01: Look, we must do ten chords.
MA01: We must do ten *baggies* and each one could have from one to ten.
MU01: From one to ten [simultaneously with MA01]
MU01: Yes.
MA01: And the duration is per *baggie*.
MU01: Yes, per *baggie*, per *baggie*.
MA02: Yes, and also the notes of each *baggie*.
MA01: We could have *baggies* of 5 or *baggies* of 10. Right? Anything [smile]
MA02: Yes. And that tells you how many fingers you are going to use, that means, how many pitches. And then, with those pitches, we see which of these [note values] is going to be [while she is pointing to the list of Fig. 4.4]
MU01: Yes, which note value is it going to have [the *baggie*]
MU02: Yes. And that is the reason why we have 10 notes, because we have 10 fingers.

The other participants immediately adopt the term "baggie" and they use it during the rest of the experience. On the same page where the musician draws the ellipse with the four marks (Fig. 4.5), MA01 adds seven more ellipses to represent the *baggies*

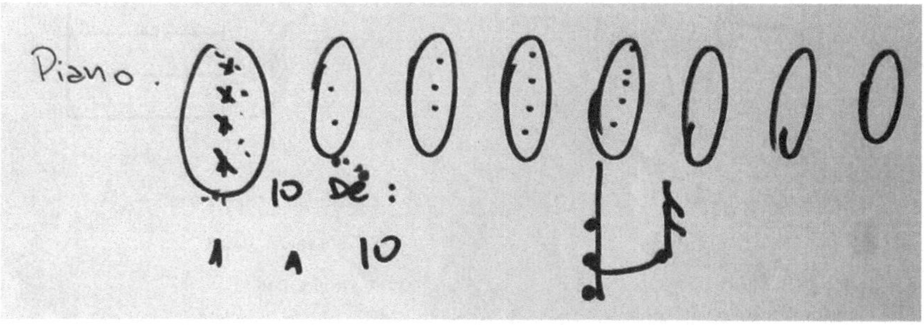

Fig. 4.5 Ellipses drawn by MU01 and MA01, representing *baggies* with different numbers of notes between 1 and 10

they are going to do for the composition. Inside each one, MA01 marks different numbers of points, exemplifying numbers of randomly chosen notes.

When the structure of the object referred to as a *baggie* was clear, MA01 perceives that the pitches will always be in the same octave. He expresses his perception by pretending to play a piano with both hands together in front of him and he warns it is going to be 10 notes from 12, reopening the debate about the number of notes per *baggie*. Thus, they modify it into two baggies of five notes—one *baggie* for each hand—so both hands could play on different octaves. This organization makes more sense to MA01, who simulates playing a piano by moving both hands in opposite directions.

Finally, the information of a *baggie* consists of five random variables: the number of digits (0–5), note values (0–9), pitch for each finger (0–11), and octave per *baggie* (1–5). The mathematics students make a visual representation of a *baggie*, which indicates all the information needed to define the piano chords played with one hand (Fig. 4.6).

Given these options, the mathematics students decide that the probability model could no longer rely on coin tosses but dice. For each random variable of the baggie (K, F, A and O), they would need a different dice, where its number of faces depends on the number of options, ensuring equal probability of success for each value. For example, the F random variable of note values requires a 10-sided dice with results between 0 and 9. They use a spreadsheet to simulate them. The mathematicians also note that the lists of *baggies* for the hands are independent of each other, so it may happen that the duration of one list could be longer than the other one. This last statement is clear for the musicians, who do not see a problem if the pianist only plays with one hand at the end.

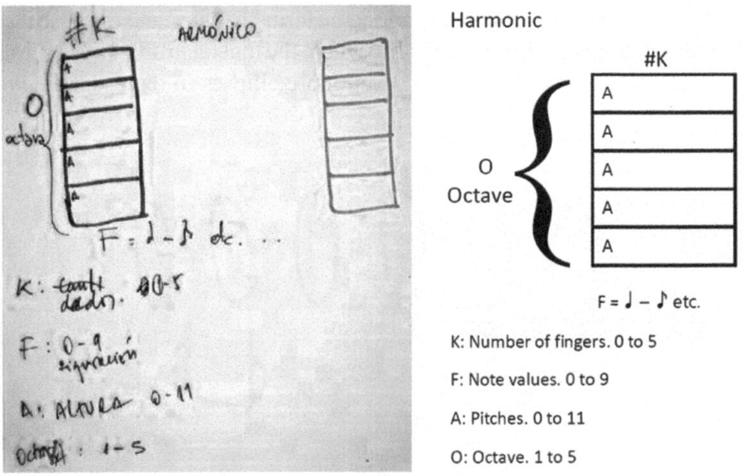

Fig. 4.6 Students visual representation of information of a baggie

4.4 Results

In commognitive research, the discourse is the primary unit of analysis (Sfard, 2008). The situations described above could be associated with a particular discourse, whether mathematical or musical, so a study from the outsider's and an insider's perspective was conducted. The first case was approached considering the line graph as a visual mediator from a mathematical discourse. Therefore, the mathematician gives the perspective of the insider, while the outsider corresponds to the musician. However, it is important to note that when the musician performs his musical interpretation of the graphics, the mathematician becomes an outsider, trying to follow the musician's explanation. As for the case of chords and *baggies*, the insider's perspective corresponds to the musician, considering chords as a keyword of a musical discourse.

Each experiment was analysed from videos and audio recordings, paying attention not only to verbal language but also to gesticulation, written records and resources. To gain evidence of the insider and outsider perspectives, interviews of participants based on recordings of critical moments were used. It should be noted that after reviewing the videos, setting up a meeting with each participant took more than five months, so having the video during the interview was important to stimulate recalls.

4.4.1 About Line Graphs and Music Reading

During the first experiment, the line graphs became a meeting point between the participants and established the tone in the composition process. The musician interpreted this visual mediator, which is characteristic of a mathematical discourse, considering a discourse that was more familiar to him. He associated this mathematical mediator with one of his practices: the reading of sheet music. On a sheet music, every note is displayed from its duration and pitch. The note value represent the duration, and its location on the staff, the pitch. Two dimensions, ordinal and intervallic, determine the pitch. Ordinal indicates that the notes are ordered from the top to the bottom, from high to low; intervallic, the distance between the notes reflects the difference of tones (Galera & Tejada, 2010; Schön & Mireille, 2002).

The musician's utterances about the graphs were only associated with the shape of the polygonal curve, he did not worry about the information provided by the axes. This was more evident when the musician mentioned that the graph could be *inverted* without affecting the mathematical relation. This practice of *inversion* could have been accepted, considering that the participants defined their process according to a *melodically interesting* graph, and for that, it was enough to observe its form, but could also depend on how faithful to the graph information the participants would like to be. Although this idea was not used, a consideration emerged from it: the interpretation of a discourse in terms of another brings with it features of the other discourse that might not correspond or even be contradictory to the initial one, "like

Trojan horses that enter discourses with hidden armies of unhelpful entailments" (Sfard, 2008, p. 35).

Another important factor in this integration process was the technology. To use technology for teaching and learning statistics and probability allows focusing an activity on the conceptual analysis rather than numerical calculation. Different methods of analysis and interpretation of large data sets are endorsed, such as simulations of random experiments, generations of different representations and automation of calculations and procedures (Chance, Ben-Zvi, Garfield, & Medina, 2007; Friel, 2007). The management of the spreadsheet by both participants characterised this experience. Having this tool made it possible to search the melodically interesting graphic, but also to simulate the experiment looking for other results beyond the heads and tails result.

4.4.2 About **Baggies** *and Gestures*

During the session, chords were approached through explanations and gestures associated with playing chords on a piano. These gesticulations were simulations of playing a piano using both hands and they were performed by music and mathematics students. Even when the students talked about a pianist playing the chords, the gesticulations were as if they were "the person performing the act, rather than taking the viewpoint of an observer of the event" (McNeill, 1992, p. 119). These gestures were as important as the talking explanations. For instance, the changes in the *baggies* structure from one to two *baggies*, one for each hand, emerged from this kind of gesture. MA01 declared during the interview, "This is logical. Although one does not know much about playing the piano, one knows that the hands are normally separate. So, it called my attention that we did it from zero to ten", explaining how he made the decision to use two *baggies* instead of one.

While the participants were arguing about the number of notes each chord would have in their composition, MA01 introduced the term *baggie*. In the interview, he explained that this idea came out when he associated the chord with the mathematics notion of set. Based on his interpretation, a chord was a set of notes and the clearest way to explain this to the musicians was using the term *baggie*. Therefore, the student used it when he associated the musical discourse on chords with a mathematical discourse that he knew, set theory, where "baggie" only facilitated communication with the musicians.

The *baggie* metaphor helped the students simplify the notion of chord, stripping it of any other characteristic as a musical object, which was not relevant at that moment, focusing their attention on what was under discussion: the notes of a piano chord. Even some of the ideas from the beginning of the discussion were mentioned again, but now the conversation seemed to be more fluent, and all ideas converged to a same goal. These students' conversation was interpreted as the emerging of a *type of communication*. It was possible to recognize new actions and reactions around some keywords, "baggie" being the most important. So, the *discourse about baggies* is about an organisation of all the random variables needed to define a random chord: the number of notes (or fingers), pitches, duration and octave. With all this information, it was possible to write chords on a staff.

In addition, the students developed a visual representation of a *baggie* (Fig. 4.6), showing all the random variables and their options. This type of communication made all the participants *insiders* of this new discourse. However, this role was not permanent throughout the activity. Participants returned to their roles of *insiders* and *outsiders* on several occasions. For example, when it was necessary to transcribe the information of the *baggies* into chords on a staff, only the musicians performed this task while the mathematicians were responsible for obtaining the random data for 10 *baggies*.

4.5 Discussion

This study delves into the actions of students while participating in an experience of interdisciplinary collaboration between music and mathematics at the university level. The actions associated with an *integration process* correspond to those where *outsider* subjects made an effort to make sense of the dialogue in terms of a discourse that was more familiar to them. For the two experiences described here, this interpretation occurred when the *outsiders* recognised in a visual mediator or in a keyword something that reminded them of one of their practices or discourses. However, the integration process could only continue when the *insiders* accept this interpretation. From that moment, the roles of *insider* and *outsider* became less evident in the conversation, and at some moments, all were *insiders* of this new type of communication that emerged from the same participants.

The distinction between the *insider* and *outsider* perspectives seems more evident when it comes to individuals belonging to different areas of knowledge, as it was in these cases, with music and mathematics students. However, at the school level, this distinction is not so obvious. Then, the question is, what happens when individuals whose specialisation is not given by any professional training carry out the challenge of integrating discourses?

The *integration* term is a contribution to the Commognitive Theory. The aim of incorporating this keyword to the theory is to analyse interdisciplinary activities in mathematics education, from the perspective of the development of new discourses. This is a first approach and new research is required to advance its development, considering these and other areas of specialisation. As the definition of integration indicates, this process is associated with a specific goal, that is, the new discourse arises from a particular need and emerges to respond to it. Therefore, the participants can develop and expand its field of action, according to their interest and objectives. This is something that was not addressed in these experiences, but it would be worth analysing in future.

At large, including the commognitive perspective to the inquiry of students' interpersonal communication broadened the viewpoint of what interdisciplinary research means, towards the development of new knowledge. In this sense, the design and analysis of interdisciplinary activities could be associated with collaborative innovation and creation. Universities, as multidisciplinary institutions, are ideal places to perform this kind of research or practices, carrying out activities that include students from different specialisations with the aim of putting their knowledge—or discourses, in commognitive terms—into perspective, through actions that implicate practice and integration of discourses.

Acknowledgements My thanks to Professor Dr. Jorge Soto-Andrade and the students who participated in these experiences. Also to the National Doctoral Scholarship 2015 from the Training Program for Advanced Human Capital. CONICYT, Government of Chile (Folio: 21151503) and the Basal Fund PIA-CONICYT for Centres of Excellence, Project BF0003.

References

Arbonés, J., & Milrud, P. (2011). La armonía es numérica: Música y matemáticas [Harmony is numerical: Music and mathematics]. Villatuerta, Spain: RBA Coleccionables.

Boix, V. M. (2010). Learning to synthesize: The development of interdisciplinary understanding. In J. T. Klein & R. C. Frodeman (Eds.), *The Oxford handbook of interdisciplinarity*. Oxford: Oxford University Press.

Cádiz, R. (2012). La creación musical en la era postdigital [Musical creation in the post-digital era]. *AISTHESIS, 52,* 449–475.

Chance, B., Ben-Zvi, D., Garfield, J., & Medina, E. (2007). The role of technology in improving student learning of statistics. *Technology Innovations in Statistics Education, 1*(1), 1–26.

Derry, S., & Schunn, C. (2014). Interdisciplinarity: A beautiful but dangerous beast. In S. Derry, C. Schunn & M. Gernsbacher (Eds.), *Interdisciplinary collaboration: An emerging cognitive science* (pp. xiii–xx). New York: Psychology Press.

Ferreira, G. M. (2001). When art, science and technology meet: The strength, the dilemma or the curse of electroacoustic music? In *Proceedings of the International Conference: Music Without Walls? Music Without Instruments?* Montfort University.

Finale [Computer software]. (2011). Eden Prairie, MN: MakeMusic.

Friel, S. (2007). The research frontier: Where technology interacts with the teaching and learning of data analysis and statistics. In G. W. Blume & M. K. Heid (Eds.), *Research on technology and the teaching and learning of mathematics: Cases and Perspectives* (Vol. 2, pp. 279–331). Greenwich, CT: Information Age Publishing Inc.

Galera, M., & Tejada, J. (2010). Editores de partitura y procesos implicados en la lectura musical [Score editor programs and processes involved in music reading]. *Revista Electrónica de LEEME [Lista Europea Electrónica de Música en la Educación], 25,* 65–75.

Klein, J. T. (2010). A taxonomy of interdisciplinarity. In R. Frodeman, J. T. Klein, & C. Mitcham (Eds.), *The Oxford handbook of interdisciplinarity* (pp. 15–30). Oxford: Oxford University Press.

McNeill, D. (1992). *Hand and mind: What gestures reveal about thought*. Chicago: University of Chicago Press.

Newell, W. H. (2001). A theory of interdisciplinary studies. *Issues in Integrative Studies, 19,* 1–25.

Noll, T., & Peck, R. (2007). Welcome. *Journal of Mathematics and Music, 1*(1), 1–6.

Polansky, L., Barnett, A., & Winter, M. (2011). A few more words about James Tenney: Dissonant counterpoint and statistical feedback. *Journal of Mathematics and Music, 5*(2), 63–82.

Schön, D., & Mireille, B. (2002). Processing pitch and duration in music reading: A RT-ERP study. *Neuropsychologia, 40,* 868–878.

Sfard, A. (2008). *Thinking as communicating: Human development, the growth of discourses, and mathematizing*. New York: Cambridge University Press.
Sfard, A. (2012). Introduction: Developing mathematical discourse—Some insights from communicational research. *International Journal of Educational Research, 51,* 1–9.
Sibelius [Computer software]. (2011). Burlington, MA: Avid Technology.
Vercoe, B., & Fiftch, J. (2005). *Csound [Computer software]*. Cambridge, MA: Media Lab, M.I.T.
Xenakis, I. (1954). Metastasis [Recorded by Jean-Etienne Marie]. On *Xenakis: Metastasis/pithoprakta/eonta* [CD]. France: Studios de l'O.R.T.F. (1965).
Xenakis, I. (1963/1992). *Formalized music: Thought and mathematics in composition* (Sharon Kanach, compilation and edition). Stuyvesant, NY: Pendragon Press.

Chapter 5
Challenges and Opportunities for a STEM Interdisciplinary Agenda

Russell Tytler, Gaye Williams, Linda Hobbs and Judy Anderson

Abstract There are increasing calls for the teaching of STEM within inter-disciplinary settings, as a way of engaging students in authentic tasks and innovation. However there have been concerns raised about the impact of inter-disciplinary curricula on mathematics learning particularly, with a concomitant need to conceptualise how mathematics might productively interact with other disciplines in STEM settings. This chapter explores cases of interdisciplinary STEM activity that arose as part of two major Australian STEM professional learning initiatives. It focuses on the variety of curriculum structures that occurred, the challenges for schools and teachers in implementing such structures, and teacher perceptions of their experiences including student engagement. Cases of inter-disciplinary tasks/investigations are presented to explore the different ways in which mathematics is transacted, and to develop a set of principles that should govern the inclusion of mathematics in inter-disciplinary settings. The cases show evidence of increased engagement and enthusiasm of students for STEM project and investigative work, but indicate the challenge for teachers of generating productive and coherent mathematics learning in inter-disciplinary settings. The results also point to institutional and systemic barriers to the wider take-up of interdisciplinary STEM activities.

Keywords Interdisciplinary · STEM · Teacher learning · Curriculum · Project based · Mathematics within STEM

R. Tytler (✉) · G. Williams · L. Hobbs
Deakin University, Burwood, Australia
e-mail: russell.tytler@deakin.edu.au

G. Williams
e-mail: gaye.williams@deakin.edu.au

L. Hobbs
e-mail: l.hobbs@deakin.edu.au

J. Anderson
University of Sydney, Sydney, Australia
e-mail: judy.anderson@sydney.edu.au

B. Doig et al. (eds.), *Interdisciplinary Mathematics Education*,
ICME-13 Monographs, https://doi.org/10.1007/978-3-030-11066-6_5

5.1 Introduction

Increasingly the acronym Science, Technology, Engineering, and Mathematics (STEM) has been associated with high-level policy advocacy across the globe. Australia is no exception to this (National Council, 2015; Office of the Chief Scientist, 2012, 2014, 2016), where concern with STEM participation and performance has driven considerable governmental and media attention. A recent comparison of STEM policy and participation across 26 countries (Freeman, Marginson, & Tytler, 2015; Marginson, Tytler, Freeman, & Roberts, 2013) commissioned by the Australian government with an intention to 'policy-borrow' from best practice, showed high level policy concern with STEM Education and with STEM research and development. That report articulated a relative lack of coordination of curricula and teacher development in STEM education in Australia, a declining comparative performance on these tests, and declining uptake of post compulsory science and mathematics (especially higher level senior school mathematics). The report identified the difficulty with definitions of STEM, which sometimes but not always included the health sciences, agriculture and/or architecture. In schools, STEM is predominantly the province of mathematics and science, and technology subjects.

Increasingly, also, the acronym STEM has shifted from being associated with the particular collection of disciplines—S, T, E and M—to advocacy of inter-disciplinary curriculum practices built around authentic problems which involve some or all of these subjects (Tytler, Swanson, & Appelbaum, 2015a). This shift has been evident in other countries, particularly the US, for some time (Bybee, 2013), and is associated with a number of separate strands of justification. First, the association of STEM subjects with national wealth producing agendas (Marginson et al., 2013; Office of the Chief Scientist, 2012) has led to advocacy of a focus on critical and creative thinking, problem solving, and digital literacy, as drivers of innovation in industry and the skill-set students need in a contemporary technological society (Tytler et al., 2015b). This is reflected in the increasing emphasis on inquiry, problem solving, and creativity in STEM curricula, particularly in high performing PISA (Programme for International Student Assessment) countries (Marginson et al., 2013). Second, and allied to this, there have been arguments for greater inclusion of engineering and technology in the curriculum, with a focus on design thinking and problem solving. Third, it is argued that a focus on inter-disciplinary problem solving reflects the way that STEM is practised in the world of work and research, and that a curriculum focus that students see as tackling 'real' problems is needed to engage them in STEM subjects and authentic STEM thinking, to stop the slide in participation in these subjects.

This advocacy of interdisciplinary STEM curricula at its core represents a critique of traditional science and mathematics curricula in their capacity to engage students in the critical and creative thinking and working, and the building of dispositions towards STEM subjects, that will prepare them for productive futures. A substantial body of research shows that many students develop increasingly negative attitudes to school science and mathematics across the primary and early secondary school years (Boaler, 1997; Goodrum, Hackling, & Rennie, 2001; Nardi & Steward, 2003; Tytler, Osborne, Williams, Tytler, & Cripps Clark, 2008). Accordingly, there has been an increasing focus on student engagement with science and mathematics, and the development of dispositions towards STEM knowledge and perspectives more generally (Breiner, Johnson, Harkness, & Koelher, 2012; Bybee, 2013; Hackling, Murcia, West, & Anderson, 2013; Honey, Pearson, & Schweingruber, 2014; Tytler, 2007). In Australia there has been increasing recognition of the importance of STEM thinking and skills for all students and of the need to bring school science and mathematics closer to the way science and mathematics are practised in contemporary settings (Hackling et al., 2013; Tytler, 2007; Tytler et al., 2008). In this chapter we explore the nature of interdisciplinary STEM and its potential to engage students in significant mathematics learning and in doing so engender positive dispositions.

Bybee (2013) has described a variety of arrangements by which inter-disciplinary STEM curriculum is implemented, pointing to a current state of confusion as to what might constitute a productive approach. Vasquez (2015) identifies different models of integration in which students learn through tasks that approximate STEM practices. However, serious questions have been raised about the capacity of integrated STEM models to support significant disciplinary learning in mathematics or science. Clarke (2014) points to the very different epistemic practices that constitute the individual STEM disciplines, in terms of the nature of their discursive practices, the type of reasoning through evidence, and the artefacts used and produced. He raises the question of how disciplinary constructs might be transformed in crossing disciplinary boundaries, and whether STEM knowledge and practice can be developed that are separate from their disciplinary antecedents. Lehrer (2016, 2017) argues that many inter-disciplinary STEM projects, while engaging for students, lack a sense of a coherent curriculum agenda and fail to engage students in the thinking characteristic of deeper disciplinary practices. They amount instead to an 'epistemic stew'. A major review of integrated STEM curricula in the US (Honey et al., 2014) found evidence of improved attitudes, but little evidence of improved learning in science or, particularly, mathematics, with mathematics teachers concerned about the level of mathematical thinking represented in these projects. A common problem with integrating mathematics into design tasks is that mathematics often plays a service role, with already-known processes used as a tool (such as calculations, or graphical work), without opportunities for the development of new mathematical insights through, for instance, students making decisions within an unfamiliar challenging problem (Barnes, 2000). Allied to these issues, there is a history of studies into integrated curriculum projects that point out the difficulty of establishing such activity within school teaching and learning cultures strongly focussed around discipline-based subjects (Venville, Wallace, Rennie, & Malone, 1998).

In this chapter, we draw on two significant Australian initiatives focused on school-led innovation in STEM curricula, to explore the potentialities and challenges of pursuing mathematics learning through inter disciplinary work in STEM. The research questions we will address are:

1. What variety of curriculum arrangements occur in these initiatives, through which inter-disciplinary STEM can be productively pursued?
2. What types of drivers, and challenges exist for these schools pursuing STEM integration?
3. What principles should apply to the productive teaching and learning of mathematics within inter-disciplinary settings?

5.2 Two Australian STEM Initiatives

The initiatives through which these research questions are pursued are:

1. The STEM Teacher Enrichment Academy developed and managed by the University of Sydney.
2. The 'Successful Students-STEM' program run by Deakin University within the *Skilling the Bay* initiative in Geelong, Australia.

The *STEM Teacher Enrichment Academy*, is delivered by the Faculty of Education and Social Work at the University of Sydney in collaboration with the Faculties of Science, and Engineering and Information Technology. Associate Professor Judy Anderson is the STEM Academy Director. This program is funded philanthropically with the intention of building STEM capacity through teacher enrichment and professional development. The Academy's flagship is a one-year programme (commenced in 2014) for teams of 6 secondary (students of age 12–17) teachers (preferably in subject leadership positions) from each of 12 schools each year. It is designed to enhance teachers' knowledge of content and pedagogy, inspiring them to re-invigorate their classroom practice and improve student engagement in STEM subjects. Each school includes up to two teachers within each of the STEM disciplines of mathematics, science and technology. This three phases program commences with a three-day residential workshop at the University of Sydney, followed by STEM teams returning to schools (supported by mentors), and concludes with a two-day workshop back at the university. An on-line platform is available to facilitate discussion and sharing of resources between teachers across schools during the program. The three-day residential program includes sessions facilitated by the university's academic specialists and STEM leaders, and previous academy-teacher-led, and peer-led sessions. The focus shifts over the three days from in-depth presentation of content and pedagogy in the separate disciplines, to sessions in which practical models of inter-disciplinary STEM work are presented and discussed, to each cross-disciplinary school team working together to develop inquiry-based learning approaches to teaching both within their subject discipline (Maaß & Artigue, 2013) as well as across the subject disciplines (Campbell & Jobling, 2014; Vasquez 2015). It culminates in school-based STEM teams planning a STEM approach to meet the needs of their students, that considers the expertise of teachers and constraints of the school context, and sharing their ideas with other STEM teams (which encourages networking across schools). In the second phase, STEM teams return to schools for at least two full school terms to work on developing, planning and implementing STEM strategies. They are supported by discipline-specialist professional mentors from the *STEM Teacher Enrichment Academy* during this time, who visit schools to provide support and assistance to teachers planning and implementing STEM strategies. The final two-day workshop focuses predominantly on school-based STEM team presentations—sharing experiences, presenting evidence of teacher and student learning, discussing issues and challenges, making closer links with other STEM teams with similar interests and/or demographics, and considering future initiatives.

The *Successful Student STEM Program* (STEM Program) is one of eleven initiatives of the Skilling the Bay, a government-funded initiative established in response to the changing economic climate in which major manufacturing industries in Geelong, Australia, have closed down. The region is working to stimulate a transition to a new, more knowledge-based economy. The STEM Program involves ten partner schools from the Geelong region, focusing explicitly on years 7 and 8. Three teachers from each of ten partner schools participate in professional development over two-and-a-half years. These teachers could be mathematics, science, or digital or design technology teachers, or teachers in positions of leadership who can support a STEM curriculum change process within the school. Teachers undergo four intensive professional development sequences, with each sequence providing two intensive days focussing on building teachers' knowledge of STEM practices and pedagogies, and a third reporting and sharing day. Schools decided their own focus for STEM improvement, which could be subject-specific innovations (e.g. focusing on mathematics or science only), innovations involving integration of subjects, or innovation across a suite of subjects that promote particular STEM pedagogies (such as design-based learning). The intention is that teachers focus on their own development, but increasingly act as change agents in their school to lead sustainable STEM innovation. A Deakin University Project Officer works with schools to support their developing practice and a Secondary STEM Teacher Network has been established.

The two programs involve similar overall purposes and structures. Both programs provide support for teachers planning and working across disciplines, but there is also the possibility of single-subject innovations. For the STEM Academy workshops, although the intention was to pay equal attention to improvement in disciplinary practice, and inter-disciplinary activity, the teachers through their enthusiasm for multi-disciplinary team discussion and planning, and commenting about the value of sharing of interdisciplinary projects across schools, triggered a change in emphasis in workshop organization. Almost all the innovations involved interdisciplinary activity, with a variety of approaches. This was also true for the STEM Program. We will use this variety of approaches to STEM implementation to examine the role of mathematics in inter-disciplinary settings, first with an overview of the variety of models, and second with case studies to examine three different models of mathematics teaching and learning within an inter-disciplinary setting.

5.3 Scoping the Nature of STEM Innovation

The regular reports required of schools in both projects, reporting and sharing events, field notes taken during professional development days, and de-briefing conversations with mentors, and, or, project officers, and surveys and selected interviews of teachers and students, have provided the authors (each of whom is involved with one or more of the projects) with insight into the variety of types of innovation undertaken across these schools, the drivers and challenges, and the change processes involved. Bybee (2013) pointed out that there is no single agreed curriculum model of inter-disciplinary STEM, and it could be argued that the field is currently in a confused state in terms of establishing productive approaches that fulfil the promise of advocates of interdisciplinarity. Within both Australian programs there was a wide variety of approaches and experiences, across the dimensions of: curriculum arrangements between subjects; teachers involved and nature of involvement; alignment with subject curricula (in some cases the mathematics is extraneous to curriculum requirements); the length of the initiative (from single events, to long-term projects); organizational arrangements (some schools formed a STEM committee to plan activity across year levels and subjects, such as planning a digital design focus across subjects); the nature of the evolution of the innovations (some schools had been working towards inter-disciplinary STEM curricular models prior to entering the programs); and embeddedness, and, or, sustainability of the innovation.

Across the two programs there was a variety of curriculum arrangements for pursuing inter-disciplinary STEM. These arrangements can be grouped into five broad but distinct models.

1. *Cross-disciplinary activities within a single subject*
   There were cases of individual subjects incorporating tasks or design work around 'real world' problems that involved one or more other STEM disciplines. For instance technology design and mathematics might be incorporated into a science unit in a deliberate way, or technology projects designed to include mathematics and science thinking. In the third case below, mathematics teachers incorporated design and science work into a mathematics unit in a developing program aimed to make mathematics more authentic and relevant. In this they were able to draw on their experience as teachers of science.
2. *A project based activity, often a design task, predominantly centred in one subject with related work, involving team teaching, taking place in the other subjects' curriculum time*
   This could involve for instance a technology class engaged in a design activity, with some of the mathematics or science needed developed separately in those classes.
3. *An inter-disciplinary project based task with teachers from different subjects planning and teaching together*
   In this very common model a cross-disciplinary team is responsible for planning a project to which two or more of the STEM subjects contribute. Examples of such projects included the design of a grandstand involving technology and

engineering design and mathematical work around quantity of material and calculations of water run-off and collection needed for gardens, short or long-term design challenges around model cars, an energy efficient house, or a robotics program involving coordinate geometry and working mathematically.

4. *Special STEM project activities*
   These are special events such as robotics days supported by university engineering students, solar boat challenge days, or visits to local university STEM facilities.
5. *A separate integrated STEM unit specifically designed to be interdisciplinary, with teachers from different subjects contributing*
   A number of schools have planned or are planning separate STEM units, often as electives, that involve the different disciplines.

5.4 The Process of Change

An analysis of reports from the twelve STEM Academy schools, and surveys administered to teachers from these STEM schools, yielded the following themes related to change processes.

5.4.1 An Increasing Focus on Authentic, Inter-disciplinary Activity

Teachers over the year became increasingly interested in, and committed to, interdisciplinary project-based learning, and confident in their assessment that this approach 'works' to engage students. Teachers worked in interdisciplinary teams to plan and became increasingly familiar with approaches in the other STEM disciplines. Some teachers (of mathematics in particular) expressed concern that these projects could compromise the integrity of mathematics. While these concerns seemed to diminish as time went on, they remained an issue for some teachers. Such concerns were in some instances exacerbated by an inability to identify where mathematics was embedded in these projects. Evidence of this came from mentor interviews and from observations of groups in workshops. In some cases, teachers of mathematics became more aware of where mathematical opportunities existed within projects over time, evidenced in presentations in the final Academy workshop. In addition, some teachers learnt to better choose projects involving mathematics that students could access and creatively explore and so deepen their mathematical understanding. Often such projects were identified through exchange of ideas between STEM teams from different schools. It is crucial that support be provided to develop such expertise in teachers, if they are to strategically represent generative mathematical thinking within inter-disciplinary projects.

5.4.2 Growing Confidence with Group-Based, Student-Centred Pedagogies

Teachers expressed growing confidence with student centred problem based learning, which under-pinned most projects, and with the value of group work around authentic problems perceived to be relevant. There was a shift towards provision of hands on exploration, choice, and open-ended questioning. There were indications of an expanding pedagogical range developed through increased interactions with teachers in other disciplines, and the demands of the tasks. It seemed that enthusiasm for inter-disciplinary activity was strongly associated with changed pedagogies and 'real-life' problems that led to increased student engagement with the tasks.

5.4.3 Professional Learning Through Interactions with 'Other' Such Learners

The major sources of professional learning reported by teachers were the ideas and experiences of *other schools* who had similarly grappled with inter-disciplinary innovations, learning from other subject teachers within their school, learning from their own experience in experimenting with implementing more student-centred approaches, and supporting student learning in settings in which the mathematics was linked to the 'real-world'. This points to the value of providing professional learning opportunities where teachers engage with 'similar others' in well-designed settings rather than predominantly follow the guidance of 'expert others' (Vygotsky, 1978).

5.4.4 Collaborative Planning and Implementing of Projects

The forming of teams and the developing of a coherent approach to STEM were crucially important aspects of innovation. In some schools the achievement of a shared purpose proved very difficult, and in all schools a process of communicating and collaborating across and beyond the team required strategic effort. Achieving buy-in from other busy teachers who did not share the same commitment, and also the wider school community, required a managed process such as providing 'tastes' of activities in a non-threatening way and/or running a project where teachers from a discipline with limited commitment in the first year were subsequently enticed through seeing the interest and learning developed by students.

5.5 Case Studies of Mathematics Within Inter-disciplinary Activity

The three case studies below were chosen from more than 20 cases within the two programs, to illustrate different ways in which mathematics was included in interdisciplinary settings. The first two cases, from the STEM Academy program, were the subject of in-depth case study exploration involving site visits, observations, interviews, and collection of student artefacts. Between them they represent a variety of the dimensions listed above, and different curriculum models. The school contexts are first briefly described then the main features of the multi-disciplinary activity in each of these schools is identified.

In Case 1, a curriculum program was established where each of science, technology and mathematics collaborated on themed projects in which subject content was developed. In Case 2, a technology design-led project on go-carts was supported in the separate mathematics and science subjects, without disturbing curriculum arrangements. Case 3 involved mathematics teachers with science teaching experience pursuing STEM projects that developed mathematics capabilities in authentic settings. The role played by mathematics, and its articulation in the curriculum, differed in each case. The challenges for teaches also differed.

5.5.1 Case 1: STEM Ed—A Collaborative Cross-Subject Program

The first case is a small independent K–12 co-educational school on the outskirts of a large metropolitan centre. Enrolment is approximately 450 students of which, 2% are indigenous. Forty full-time teaching staff work across the year levels with single classes in each of the primary years and usually only two classes in each of the secondary years. Due to falling enrolments in senior secondary higher-level mathematics and science, and students' poor attitudes to mathematics in particular, the Deputy Principal (who leads and teaches mathematics in the school) determined the school needed to adopt new approaches to the teaching and learning of the STEM subjects. Prior to participating in the STEM Academy, and in collaboration with the head of technology and a member of the science faculty, he trialled and evaluated a STEM project with a small group of lower secondary students in 2015. The school also funded the development of a new learning space which students in year levels 7 and 8 (12–13 year olds) co-designed to accommodate an integrated approach to STEM teaching and learning.

With the support from the STEM Academy program, and encouraged by positive responses from students, parents and other staff members, the initial project work was expanded in 2016 to all year 7 and 8 level students. They followed the same scope and sequence of learning activities as undertaken by their predecessors the previous year. The small school size limited timetable flexibility so creative ways to set up an inter-disciplinary programme were explored. 'STEM-Ed' 'mathematics group' 'science group' and 'tech group' was the terminology used by the school for their program, which runs within usual classes in each discipline. Substantial re-design of the curriculum based on syllabus requirements for science, mathematics, and mandatory technology was required. The STEM team based this around 'themes' from the science syllabus document with one theme as focus in each of the four school terms throughout the year (e.g. The Rocket Project). The mathematics and mandatory technology syllabi requirements were then mapped onto these themes as far as was possible. However, time was left available in the mathematics discipline classes to teach mathematics topics which did not align with any of the identified themes. The school thus met most science and technology curriculum requirements through STEM-Ed with more of the mathematics curriculum taught external to STEM-Ed. Project briefs and assessment rubrics were designed so that students had clearly defined intended outcomes for each project, along with assessment criteria, and the expected standards. Each project brief began with a challenge then clarified the technology, mathematics and science the students were required to incorporate into their designs. Each project contributed a mark towards the final grades for the students in each STEM subject.

Students worked in small teams for 8 weeks within their separate discipline groups. It was not unusual for a teacher from one discipline to visit their class as they worked within another discipline and to take interest in what the students were achieving.

STEM-Ed mathematics group members (students) undertook particular roles within each project as well as developing ideas together. These roles included researcher, PowerPoint presentation developer and 'Think Bigger' person, who explored an idea of their own choice beyond but related to the task description. There was thus scope for autonomous and more substantial mathematical engagement in the Think Bigger aspect of the project at least. As each group presented their findings to the class, there was also opportunity for the class to learn from each group.

In the Rocket Theme Project, the mathematics component was associated with the identified theme but ran for the most part separately to the technology and science components of the project. The STEM-Ed technology group teams built and tested a rocket and in doing so, they tested possible materials as part of their STEM-Ed science group. The STEM-Ed mathematics group researched time on Mars: comparing Mars time with Earth time, considering months and seasons, and designing a school day. Two of the 'think bigger' tasks students explored were: (a) leap years on Mars, and (b) using space station data to find how long it takes for light to travel from Earth to Mars in light minutes to determine how long it would take to communicate between the two planets with text-like messages. Figure 5.1 shows some of the work undertaken by one Year 8 think bigger student who had not previously been aware of the 'Right Angles in Semi-Circles Theorem' or trigonometry.

In communicating to the teacher that he had found the distance from Earth (blue ball centre right) to Mars (at L3) (through research), and now needed to find the lengths from the satellite (at L4) to Earth and the satellite to Mars, the teacher suggested researching right angles in semi-circles, and trigonometry to see whether

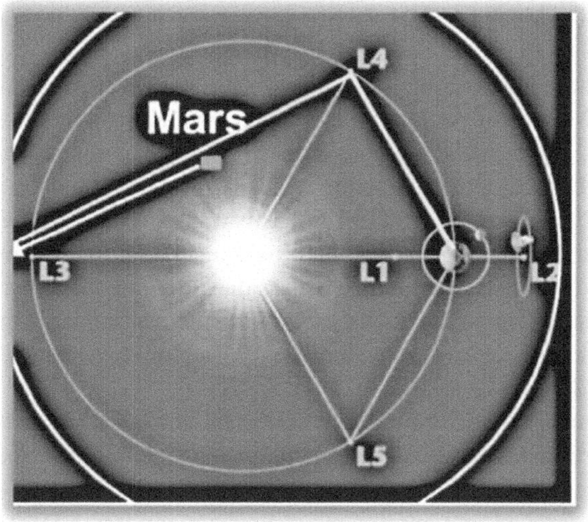

Fig. 5.1 'Think Bigger' student diagram used to work out time taken to communicate from Earth to Mars

he could find anything useful. By first encountering these mathematical ideas at a time that was useful for him, the mathematics was meaningful on first encounter. The student used this mathematics successfully to find a solution to his problem. This provides an example of a student who found a way to find some information and identified what more he needed to know to find out what he needed to know. Krutetskii (1976) identified knowing what information you still need to find as an activity undertaken by highly capable student mathematicians.

The six students interviewed (three girls and three boys) were asked questions about: how learning in STEM-Ed compared to learning in other subject areas, what they had learnt, and their career aspirations at the end of primary school and now. All students considered they had benefitted from STEM-Ed and listed this 'subject' as either their favourite or one of their favourite subjects. This included students (both girls and boys) who had been more interested in art and humanities at the end of primary school.

> When you're in STEM, you learn the information, then you get to put it into a practical use in tech. Also in science we get to do field tests of what we've learned, and the same with maths. It's just very interesting.
>
> In science in primary, it was like you need to learn this, if it's not that, you're wrong, but now it's more like you get to choose, you have more freedom and creativity. (STEM-Ed students)

As the principal was supportive and provided funds for the initial development of STEM-Ed before it was implemented in the school, and the deputy principal STEM-Ed mathematics team leader, was responsible for the school time-table, many of the usual problems with organisation did not arise in this school. A session for parents was arranged when the idea of STEM-Ed was being developed in 2013 so an interested and supportive broader school community was fostered.

Time and collaborative communication were considered essential for the development of STEM Ed. Key resources required for scaling-up were funding for collaborative time and professional learning of teachers, both of which were provided by the STEM Academy. Committing to the Academy also increased the perceived obligation of the school to progress STEM-Ed. This was extremely useful for increasing the participation of more teachers in the program. In addition, the mentor reported supporting Academy mathematics teachers as they identified key components of the mathematics syllabus associated with each of the themes and supporting their decision to leave other aspects of the mathematics curriculum to be taught independently of the STEM-Ed project work.

While it is too soon to determine whether the initiative has had an impact on post-compulsory enrolments in higher level mathematics and science, there is evidence from school data and interviews with teachers and students that attitudes to the STEM subjects have improved and enrolments in upper secondary STEM are improving for 2017. Sustainability of the initiative is imperative if this work is to continue to have an impact. The Deputy Principal was a key driver of this initiative but he is moving to another school in 2017. While there is considerable momentum in the school around the STEM-Ed approach with two of the original three teachers who developed the STEM-Ed program leaving the school, it will be critical for the head of technology to keep the momentum going. He is aware of this responsibility and is already considering which staff he will be able to draw upon to sustain the momentum, and who he will need to support as they develop their STEM-Ed skills further. He envisages a very collaborative STEM-Ed team. Plans are already in place to extend the program in 2017 into the primary school years and to introduce a Tinkerspace for the younger students. For this particular school, being small has also assisted in the expansion of the project beyond one class of students in 2015 to all students in years 7 and 8 (Age 12–13: a total of 5 classes). Executive leader support and a supportive parent community have been essential to the growth and maintenance of the initiative. Some of these characteristics contrast with the second case study school.

The STEM-Ed program at the school seemed to be a fortuitous combination of circumstances—the drive of the leadership team and energetic mathematics and technology coordinators, and the STEM Academy support, served to enlist the support of teachers and parents. Drive was needed to overcome institutional barriers such as timetabling and curriculum constraints. The question mark against a sustained STEM focus serves to underline the challenges faced in establishing a cross-disciplinary innovation of this type.

This case study shows that the structure of the 'think big' aspect of the responses to each project gives opportunities to select and explore new mathematical ideas and the nature of the task is critical to the degree to which this is possible (Kieran et al., 2008; Williams, 2002). Students require opportunities to explore unfamiliar mathematical ideas by *recognizing* the relevance of known mathematics, *building-with* it in unfamiliar sequences and combinations, and synthesising these (*constructing*) to realise something mathematically profound (Dreyfus, Hershkowitz, & Schwarz, 2001; Williams, 2007, 2014). Key to achieving this is teacher awareness of the mathematics embedded within a STEM activity. Such activity was not occurring in all STEM-Ed projects undertaken by the maths group. For example, the Garden Project undertaken in STEM-Ed included far more revision of known mathematics (recognising shapes and their properties) than opportunities to use known mathematics to solve unfamiliar problems. One mathematics teacher emphasised the difficulty of 'tying' the mathematics into inter-disciplinary topics, and the need for support to do this:

> I'm always looking out for where the maths is and then trying to tie it all back in, which is a challenge for myself. It's always hard when you're teaching indices or something where you go here's the rule, just do it, versus can we actually apply this to something that we're doing … I'm finding it quite challenging, but I'm enjoying the challenge … [the Vice Principal] has got a lot of maths experience and he's the one that can just whip out all these connections. So I feel supported. (STEM-Ed Mathematics teacher)

A key aspect of some mathematics teachers' experience in STEM Ed was therefore the challenge of extracting meaningful mathematics from these cross disciplinary topics in ways that can extend students' mathematical thinking. This quote implies the importance of mathematics 'experience' in developing the flexibility needed to make the connections that generate mathematical ways of thinking and working, beyond the procedural traditions of the curriculum.

5.5.2 Case 2: Whole of Level Design Technology-Led STEM

Unlike the first case, this case involves mathematics and science teachers supporting a Design Technology class project within mathematics classes, and occasionally becoming involved in team teaching. The mathematics was developed at different levels to match the differentiated skills of the students.

The case is a medium sized Catholic systemic co-educational secondary school located in a large provincial town with about 860 students of which 3% are indigenous and 2% have non-English speaking backgrounds. Many of the 70 teaching staff have taught at the school for some time, are very experienced and engaged in a range of ways with the local community. Similar to the first case study school, the motivation for introducing a STEM education initiative was to address the falling numbers of students doing higher level mathematics and science subjects in the senior years but of particular concern was the lower representation of female students in the STEM subjects.

Before engagement with the STEM Teacher Enrichment Academy program, there had in general been little collaborative curriculum design work in the school, including between teachers of the STEM subjects. The Academy provided the much-needed impetus to begin their journey and at a pre-Academy visit by an author of this chapter, it was evident that the team of STEM teachers had met on several occasions to plan an integrated STEM approach focused around the design work taking place in a year

10 elective Design and Technology class. During the Academy, the teachers further developed these ideas and benefitted from feedback and suggested approaches from Academy mentors and teachers from other schools.

During 2016, the thirty students in the Design and Technology (D&T) class taught by the leader of the STEM team in the school were to research and design a billy-cart. However, the intention was that all year 10 students would participate in the project work by exploring the associated relevant mathematics and science content and processes within regular timetabled lessons in those disciplines. It was fortuitous that students in the D&T class were spread across all of the year 10 mathematics classes and science classes and were thus able to share what was happening with the billy-carts in D&T and elicit help with the Billy-Cart Project from students in other STEM discipline classes. Thus all year 10 students took part in the design of the Billy-Carts.

The planning stage and the commitment of the STEM team in undertaking this stage throughout the Christmas holidays was crucial to the success of the Billy-Cart Project. An on-line website was constructed to have everything ready for the coming year. The overall program for the technology class was outlined and teachers of mathematics and science designed learning experiences for their classes based on what the students would require at particular times during the year. This meant that the technology program drove the scope and sequence of the mathematics and science classes during year 10.

During the billy-cart construction stages, students took the billy-cart chassis or particular parts of the billy-cart (e.g. possible wheels) to their mathematics class to discuss different features and plan for the next stages of construction. Students from the technology class were able to lead learning conversations, and others would ask questions to deepen understanding as they all worked to solve problems that were encountered along the way. In mathematics, contributions to the project ranged from basic measurement and financial mathematics to applying algebraic modelling using Geogebra. To find appropriate wheels, students explored associations between features like wheel circumference and thickness, and billy-cart speed and stability, and evaluated factors they found had an impact. This overlapped with what occurred in science classes where students studied effects of forces on billy-carts and concepts associated with motion. Thus, there was some applying of previously learnt procedures in familiar contexts that led to mathematics becoming meaningful to some students who had not previously seen its relevance. At least one student, in interview, indicated this led to reconsidering mathematics subject choices. In addition, there was some recombining and/or resequencing of previously known mathematics to solve unfamiliar problems. Insights developed from this activity were about different features of wheels and how they could affect motion and the mathematics employed was crucial to finding this out.

The mathematics teachers designed and adapted tasks to facilitate learning for their streamed classes. In ascending order, these tasks focused on: making mathematics meaningful; assessing billy-carts on safety criteria; and using digital applications to identify features of billy-carts that affect its motion.

The students who were perceived to struggle more with mathematics developed a budget for the billy-carts, made scale models using centi-cubes, sketched their models from different elevations, and designed a race-course around the school for the final event in the Billy-Cart Project. As noted by the project leader, "the students went to the Six Maps website, calculated the area and perimeter of the school, then went out and measured a course with a trundle wheel and created a 'strip map' which they then drew up". She noted the students were much more engaged than in usual mathematics lessons, spending more time on task, and if asked to do any other mathematics topics, they inquired "What's this got to do with billy-carts?" This project provided these students with opportunities to make meaning of mathematics beyond routine rules and procedures (Skemp, 1976).

The mathematically competent students investigated the mathematics behind the safety ratings applied by the Australasian New Car Assessment Program (ANCAP) then applied this mathematics to determine the safety of a late model Holden, an older car and their dream car. They sketched what they saw, marked in what they were able to measure, and worked out what mathematical information they still needed to find because it was unable to be measured, again showing a capacity to identify required information (Krutetskii, 1976). This task created discussion as students struggled to work out what they needed to know how they might be able to find out. When students found that the hypotenuse could not be shorter than the height of the triangle, informal development of ideas about the relative lengths of sides of triangles was stimulated for some students. These students had thus discovered and decided to explore a mathematical complexity (Williams, 2007).

The students that teachers perceived to be highly capable mathematically used a Geogebra animation designed by their teacher (physics teacher member of the STEM team) to explore the impact of various factors on the speed of billy-carts based on physics formulae.

While much of the project work occurred in lessons taught by the relevant subject teacher, some team teaching occurred during the year. A positive outcome from this approach was that the technology students appreciated their mathematics teacher coming to their technology class to help them with their billy-cart designs—the usefulness of mathematics became more evident and they were able to 'transfer' knowledge more readily between their STEM subjects.

Several key features of the approach at this school lead to the success of the STEM program. These included the passion, vision, and leadership of the technology co-ordinator who was the key driver in designing the Billy-Cart Project, developing a collaborative STEM team, and co-ordinating meetings with other staff members involved, and engaging the community with the project. The STEM Academy school-based team worked through the Christmas holidays to prepare everything for staff and students when they returned to school. The large number of teachers enticed into the project by the STEM team provided opportunities to source many types of expertise. Community engagement was critical to the success of this project with, for example, teacher contacts with the Vintage Car Community resulting in a Motor Show where cars from the early 20th Century through to the present day were displayed on the oval. There was also ongoing encouragement and support from the Regional Catholic Education Office who subsequently drew on the expertise this team developed.

The overall engagement of a large number of teachers from the school, not just the six teachers in the STEM Academy was crucial to the degree of success achieved and should impact on the sustainability of the program into the future. In the final report to the STEM Academy, the team leader wrote:

> A key success to the STEM project at [this school] was the involvement of the TAS (Technology), Mathematics and Assistant Science co-ordinators. This enabled effective communication within and across KLAs [Key Learning Areas]. There were 6 Science teachers involved in this project, together with 8 Mathematics teachers and 1 Design and Technology teacher. In addition, 2 staff played a significant role in the support of IT and implementation of STEM at a leadership level in the school. A total of 17 staff were involved in this pilot project, this equates to 25% of the full-time teaching staff at [the school].
>
> This school has played a leading role in stimulating interest in STEM Education in other schools in the region and beyond and providing professional learning for them. This should also impact on project sustainability.

Student interviews confirmed the success of the STEM approach which heightened their enthusiasm for the STEM subjects. The Design and Technology (D&T) students who had previously not looked forward to mathematics and science described how STEM activity had given them focus, and learning that had resulted:

> Yes, sort of like we were saying before, like when you look at the timetable for the day and it's only got three lessons on the same thing. So you walk into the day thinking like, you know, three lessons on the billy cart and how it works and stuff like that, … so you sort of get three lessons to work on one thing, rather than one hour … You've got to change it up sort of thing. So it's definitely helped me,

It also led to an appreciation of collaborative group work:

> … if you come to a problem, you've got to try and create another alternate way of fixing that and like when you're by yourself it's a bit hard. Like you try and think like, "I don't know how I can do this," but when there's two other people in the group that can combine ideas, like have a real good idea that fixes it.
>
> People know different stuff and they bring different things to the table and have more options.

Participating in STEM projects has led to some of these D&T students who had not considered doing mathematics in year 11 reconsidering their options. For example, when asked whether he had considered mid-level mathematics prior to the project, one boy stated: "No not at all!" The project had influenced this student's subject choices: "I think it's definitely helped a little actually because of Maths … [I've] been more engaged … just with the stuff we're doing, so from that—wheels and angles and stuff like that".

The energy and enthusiasm of this school for designing new and more integrated STEM approaches is still evident with plans to design a project called "Save the Earth" for year 7 students towards the end of 2016. To facilitate this, and further STEM work, the Deputy Principal who manages the timetable has been co-opted as a member of the STEM team so that school structures which were problematic or fortuitous during 2016 can be planned in 2017. These included timetabling all members of the STEM team as teachers of year 10 classes, providing timetabled sessions where STEM team members can continue to carry out professional learning for other teachers during their classes in 2016, and timetabling meeting times for the STEM team. In addition, there are provisions for the introduction of a new elective subject for year 9 and 10 students, implemented in 2017.

Opportunities for students to engage with mathematics in unfamiliar ways were increased by the STEM team's ability to think creatively in developing and extending the program and bringing other teachers on board. The passion and energy of the school-based academy STEM team spread through the broader STEM participating teachers at the school.

5.5.3 *Case 3: Engaging in Mathematics Through Within-Subject STEM Investigations*

Case 3 involved years 7–8 mathematics teachers developing STEM investigative projects within the mathematics curriculum in order to better engage students conceptually. Case 3 is a state girls' secondary college, which joined the STEM Program in order to attend to a continuing problem of students entering the school at year 7 with weak mathematics background, low aspirations towards STEM careers, and a decline in senior mathematics and science enrolments. In order to address these issues, a teaching team focussing principally on mathematics was chosen to participate in the STEM Program. The teachers' framing of STEM was largely based on a need to improve students' engagement and application of mathematical ideas, so was largely focussed on improving mathematics teaching and learning. The school endeavours to have the same teacher teaching mathematics and science at years 7 and 8, allowing for some co-ordination of the teaching of science and mathematics. Therefore, two of the activities generated by the team have scientific concepts embedded in them to frame the nature of the problem being explored mathematically.

During the first professional development session, problem solving, modelling in mathematics, and collaboration during problem solving, were introduced to the participating teachers. The teachers from School 3 were intrigued by this problem-solving approach, particularly because they had been experimenting at the school already in a year 9 mathematics elective where students are required to build a chair and explore the mathematics involved. This activity was taught by a single teacher in an elective class, but the STEM program teachers decided to expand this approach into the mainstream mathematics classes at years 7 to 9. In the first year of the STEM Program (professional learning sequence 1) the teachers developed a ramp investigation for year 7 classes. In year 2 they refined the ramp investigation, and developed two new investigations for year 8 students: a tank investigation (professional learning sequence 2) and a clothes investigation (professional learning sequence 3). The process of developing the investigations evolved over time, and refinement and redevelopment is a crucial part of ensuring the investigations are manageable for students, and that the necessary scaffolds are provided through differentiation of the tasks and provision of resources. The intention is to have one investigation per semester through years 7 to 9. The investigations last for about two weeks; any longer than this and the teachers found that the students lose momentum.

The three investigations are described in Table 5.1 structured under the headings 'Immersion', 'Mini-inquiries', and 'The big question', which drove the structure in each investigation.

The process of investigation is illustrated by the Clothes investigation in which the following steps were carried out.

The first step was posing the question. Students were encouraged to consider at what time during their school life they need the most amount of new clothes. Thinking of their parents who need to balance their finances, students are asked, wouldn't it be good to know the budget for when they need to buy the most clothes? A PowerPoint presentation was given to the class, and support materials were available on the school intranet.

Students then undertook mini-inquiries where they collected the heights of students in their school, a local primary school, as well as secondary data from the Bureau of Statistics (being a girls' school they needed data for boys). One teacher stated that when discussing the data collected from within the school, one student queried the result that the average of year 9 was taller than year 10:

> I said have another look at the data, and she said that one student had recorded themselves as 185 cm tall. And then she looked at how tall that is. And I said do you know a year 9 student who is that tall? Well, no. And then we opened up the discussion about outliers, which is probably a little bit beyond where she's at. But I did speak to some of the Further Maths teachers in year 12, and they said that it's often a conversation that kids don't have and they don't know what you mean by an outlier. But even if I've sown that seed, then when it comes up later they might be able to recall what we talked about. So, she decided to discard that bit of information.

Having examined data from primary and secondary schools, students were asked if they had enough data to answer the question. Students needed to be led to consider the predominance of female data, and some students were not willing to examine this limitation of the data. Some students did realise the need for more male data, and discussed the need for secondary data. They searched the Australian Bureau of Statistics website and found some male data and the teacher made that available for the students: "Some of them were finding it hard to get the data themselves, so I randomly selected different year levels' male data and if they wanted they could access that".

In addition, students were asked to consider data from home, for example growth marks on the door-frame showing height at different ages.

In preparing for the report, the students were asked to prepare a plan for responding to the question. The report was aimed at parents to support their budgeting for new clothes. The instructions included attention to hypothesising, arguing and concluding, communicating, and using appropriate mathematical representations. The task needed to be differentiated by limiting or opening up the scope of the problem

Table 5.1 Learning sequences for B college year 7 and 8 mathematics investigations

Phase	Year 7 ramp investigation	Year 8 tank investigation	Year 8 clothes investigation
Immersion			
Understanding the problem	Investigate different ramps for the disabled in the broader community by walking around town, including a local hospital and the school itself. *Are all ramps the same?*	Investigate the school vegetable garden watering system. Students are asked to observe how the beds are watered from rain-water tanks, as well as determining how these tanks are replenished	Investigate the growth patterns of students and determine when they are likely to need to buy the most clothes. The Big Question is introduced: *At what times during your school life do you need the most new clothes?*
Guiding tasks	In Mathematics, measure various ramps, and produce scaled drawings of ramps Investigate different ramps using the dynamic trolleys in Science class to examine how inclination affected effort In Science test how the ramp elevation affected speed of descent	Science workshops on transpiration and evaporation Groups of students are allocated individual garden beds, and the Big Question is introduced—*Are our new tanks big enough for our garden*? A brain-storming is used to identify a range of mini-inquiries to be undertaken	Students bring their own growth data from home, such as measurements of height growing up Students brainstorm as a group: – What data have we collected that will help you to answer the question? – What can we do with this data to help you answer the question? – What is the question asking? – State your hypothesis

(continued)

Table 5.1 (continued)

Phase	Year 7 ramp investigation	Year 8 tank investigation	Year 8 clothes investigation
Mini-inquiries			
Asking questions	Students undertake a series of mini-inquiries: – Compare different ramp lengths in the garden and then rank the effort needed to get a wheelbarrow up each ramp – Hands-on activities give students a physical sense of elevation versus effort – Explore the ramps around the school using a wheel chair to investigate ease of being pushed up or down, and the difficulty of wheeling oneself up or down	Students generate inquiry questions that are used to direct classroom learning such as: – How much water do our garden beds receive and use? – What is the area of the collection point (rooves of school buildings) and how much rain is collected? – What is the total area of the garden beds? – How much water is used by our garden beds? Additional mini-workshops: – what does 1 mm of rain really mean? (depth of rain recorded, and collection area leads to volume of water collected) – transpiration rates are linked to daily temperature readings	Students place sheets in public areas of the school and students put their age, and height onto the sheets. Averages calculated, outliers identified, comparisons made across year levels Each year 8 class collated and recorded the data in an excel sheet Collected height data from primary school on sheets Students decide whether there is this enough data to answer the question? (e.g. Predominance of female data) Students considered data that is kept at home, e.g. Height marks on the wall Additional data is collected from secondary sources, such as Australian Bureau of Statistics

(continued)

Table 5.1 (continued)

Phase	Year 7 ramp investigation	Year 8 tank investigation	Year 8 clothes investigation
The big question			
Analysis and conclusion	The big question is introduced—"*What would be the best ramp for wheelchair access to the deck in our Garden*?" Students design a ramp to scale and justification for their design	Groups of students work on the findings arising out of the mini-inquiries, then report solutions to the big question to rest of the class	Students make decisions about which data they will analyse, then report their findings and recommendations in a report for parents

Adapted from Hobbs, Cripps-Clark and Plant (in press)

that students reported on. For example, students decided whether they made recommendations for females, males or both. A teacher explained the importance of scaffolding the students:

> We do try and scaffold it. So, for some students, if the question is too big then maybe focus just on females, maybe reduce the amount of data as it was a bit overwhelming for some students – different ways to differentiate the task. I just speak to students individually to see how they're going and suggest where they can reduce it and give them some support in that way.

Developing a Culture of Investigation

Through these open investigations, the teachers are attempting to bring in real world mathematics where students can see the application of mathematical concepts. One teacher reported that as the investigations become more entrenched into the normal practices of the teachers, they have allowed the staff and students to see learning mathematics in a new light, not so routine or abstract. The teachers have found that students are more engaged, as the investigations relate to questions that are relevant to students' lives. In making this new approach sustainable, there is a need to bring other teachers on board and embed the practice in the school so that there is an expectation and willingness to use this approach.

The teachers reported that more mathematics staff are implementing these investigations in their classroom. Initially just the three STEM Program teachers developed and implemented them for their year 7 classes, but over time some new year 7 staff have started implementing the ramp and water investigations: "Hopefully, we are getting staff experimenting with this teaching practice and be a bit confident to go into the classroom and use these teaching methods".

Recognising the need to change is a pre-condition for changing practice; the STEM Program teachers saw the incorporation of STEM practices as a valuable conduit for promoting a new way of learning that promotes problem solving and creativity. In thinking about some of the key outcomes of being involved in the program, one teacher mentioned that "Our big push is that engineers are solving questions" and that this vision underpins their planning for these investigations. One teacher stated that these investigations are a way to bring STEM practices into their teaching, not just when completing these activities, but staff are tending to apply these teaching practices in their general teaching: "They become your 'default' position, that you don't always just have that really structured or skill-based teaching of mathematics". One teacher commented further: "It's not just about the investigations, but trying to model to other teachers that STEM practices should be part of your teaching. Makes teaching more exciting, also makes the maths more exciting".

Documenting both the activities and the process of implementation was shown to be critical to encouraging uptake of the activities by other teachers and ensuring the spirit of the pedagogy is maintained. As the activities are applied by the other mathematics teachers, further materials are being developed to guide the teachers on how to support the students in a differentiated way. For example, a graduate teacher developed a booklet to help himself understand how to run the units and then another teacher developed the booklet further. Initially the booklet was intended for the teachers but then they decided it would also be useful for students, particularly students needing additional support such as guidance in collecting and representing data. One teacher said that this type of material "embeds it a bit more" in the school, so their intention is to document the other units in a similar way to ensure the activities are not shelved.

Important factors in this sustainability are teachers recognising a need for change, and that the proposed change is effective. The teachers regularly gathered feedback from the students as they were developing the units. They have found that, for the Tank activity, for example, students appreciated the opportunity to see how mathematics related to their real life and the differentiated nature of the tasks (such as through multiple entry and exit points), but that some students still found the task difficult. Using a Plus/Minus/Interesting (PMI) framework, students reported comments such as the following:

- PLUS: "I liked how it was a real-life problem. It made me more interested in doing my maths work"; "I liked how we worked independently"; "Making the graph was fun"; "I liked working through the questions in my own time"; "I like that it was real work maths and it actually meant something to use because it is where we live".
- MINUS: "I didn't really enjoy that I didn't get to start the question because people didn't hand in their data"; "Some of the questions were very difficult"; "Some of the calculations I found hard to do".
- INTERESTING: "I found it interesting how much water Geelong used".

Such feedback was shown to be important in refining the activities and their implementation, but also in promoting these STEM teaching practices within the school more broadly. The data is being reported to other teaching staff and school leadership in order to inform the broader STEM agenda currently being developed at the school.

Developing the Tasks

Once the open investigations structure was developed and trialled by the teachers, the difficulty they then faced was in deciding which topics could be approached through an open question and how to make the problem 'relevant' for the students. The teachers also decided to focus on areas of the curriculum that were not typically addressed adequately. Two teachers explained with respect to the Ramp activity:

> This task came from one teacher saying we want to do linear relationships, so we sat down and thought how can we turn that into a question? And at the start we were struggling. But it is interesting that when you get your mind into that way of thinking it's interesting where you'll end up.
>
> This is how we've come up with the questions. We look at what we teach across the whole year. For example, for the ramp investigation, we decided we didn't do enough geometry,

didn't cover triangles that well, so look at things that we normally skim over and try to make that more in-depth.

The tasks were generally developed to be contained within the mathematics classes, but teachers who were also the science teachers could make links to concepts from other subjects, particularly science concepts in the Ramp investigation and transpiration in the Tank investigation. The real-life problems that they selected are by their nature complex in terms of knowledge needed to solve the problem, as well as in the possible solutions and mathematical thinking involved.

5.6 Discussion

These case studies have presented three different approaches to interdisciplinarity in their curriculum arrangements and the way the mathematics intersected with other subjects. In Case 3 the project sat entirely within the mathematics subject but was inter-disciplinary in the way the mathematical ideas intersected with the other disciplines in approaching the task. In each of the investigations the mathematics was central to resolving the task. Within cases 1 and 2 there was variation depending on the particular project, or on the particular mathematics class. Sometimes, as with the billy-cart wheel investigations, the mathematics was embedded, and differentiated. In other cases, the mathematics was extended into aspects extraneous to the central theme, to develop mathematical thinking unhindered by that requirement, such as with the exploration of geometric shapes associated with the garden design or the calculation of astronomical distances (Fig. 5.1). This further mathematical thinking was often exploratory. In all three cases, mathematics and other knowledges were developed side-by-side in a way that intersected with but was not totally confined by the problem, and fed off the authenticity of the problem to engender interest and further develop the mathematics. In each case also the STEM approach was introduced with the intention of enlivening the mathematics curriculum to develop more positive dispositions in students towards mathematics and STEM. There is evidence in these case studies that this manner of teaching mathematics is successful in doing this.

In each of these schools, a large portion of the mathematics curriculum was left undisturbed by the STEM sequences, more so in some cases than others. Schools have not yet focused to a large extent on building a coherent curriculum through STEM although each has made a start: Case 1 school has mapped science and technology against interdisciplinary learning at year 7 and 8 and will continue to develop this more, Case 2 school has focused attention on drawing out the curriculum links in projects they have considered to have potential, and Case 3 school has focused on identifying a project that fits with several areas of mathematics within the curriculum. It is clear that in engagement with ideas and responsiveness to mathematical ideas associated with authentic problems, that the grounding of mathematical processes within such tasks can occur in areas as diverse as motion, functions, statistics, measurement, trigonometry, descriptive geometry and geometrical theorems, and number patterns. Where there is significant mathematics involved, there are opportunities for analysis, modelling, design, evaluation, and synthesis that include requirements to transfer between representations. Situating these tasks in community settings, and providing teachers with ongoing access to the ideas of 'similar others' can be crucial as can access to strategic leaders who are willing to listen and provide the necessary support to help to overcome the challenges that STEM teams encounter. There was also evidence of teachers supporting the types of high-level mathematical thinking just described, arising from some of these tasks. There was evidence also of mentors providing valuable support for teachers to identify, interpret, and create mathematical opportunities pertinent to particular STEM tasks, implying that this way of working is challenging for teachers of mathematics who are used to working in more systematically framed mathematics curricula. This points to the importance of raising teacher expertise in identifying and embedding substantial mathematical opportunities within tasks. There was ample evidence in these and other cases of teachers' enthusiasm for grappling with these tasks, their perception of students' enthusiasm, and student reported valuing of this way of learning.

From these case studies and experience with the programs more generally, and drawing on findings from the research literature, we argue that there are a number of principles that underpin productive mathematics learning within interdisciplinary activity:

- Tasks should have an affective payoff—students need to want to do it;
- Tasks are open and emphasise problem solving that involves creative use of mathematics rather than external control of student thinking;
- Tasks encourage/entice students into using mathematics in unfamiliar ways, involving new representational moves, transformations, sequencing and combining of mathematical ideas, and synthesising ideas so there are opportunities for something mathematically profound to emerge; and that
- Where tasks do not fit with the above principles, but primarily involve using previously known mathematics as a tool for exploring authentic problems, the relevance of mathematics can be increased if it is seen to lead to fresh insights into the problem.

5.7 Conclusion

In this chapter, we have described the progress of teachers and schools in developing approaches to inter-disciplinary STEM curriculum innovation within two substantial Australian professional learning programs. We have identified the different ways in which such inter-disciplinary work can proceed, both in terms of productive approaches to the relations between teachers and STEM subjects, the ways in which mathematics teaching and learning can operate productively in an inter-disciplinary setting, and ways some of the challenges encountered have been overcome. Evidence from these programs suggests that while there are reported concerns (Honey et al., 2014) about the integrity of the mathematics curriculum and mathematics learning in such projects, this was not a significant feature of what occurred in these schools. There was some indication that teachers' capacity to develop mathematical components of projects that conformed to the principles above improved over time, aided by sharing ideas across schools (with similar others) and sometimes input from mentors (expert others). At least some of the teachers had previously experimented with inter-disciplinary STEM projects for several years in their schools. This suggests a need for such projects to extend over time to provide opportunities for teachers to be supported as they experiment to develop appropriate pedagogies and become able to identify and/or embed substantial mathematics within interdisciplinary tasks.

Enthusiasm for inter-disciplinary work increased for teachers in these schools as their students found these activities absorbing and engaging. This was true both where the activities involved teachers from a number of STEM disciplines, and also where an inter-disciplinary approach was embedded within the mathematics program. One of the key arguments therefore is that at least some of the mathematics curriculum can be learnt in an inter-disciplinary fashion to harness student enthusiasm through its depiction of mathematics as relevant in interesting and unexpected ways. In each of these cases mathematics provides a window through which patterns and structures in natural or designed systems can be quantitatively discerned and further explored.

There is a significant question, however, about the way mathematics is enlisted and used in STEM activities, and the extent to which such inter-disciplinary work can address the mathematics curriculum. This is related to our argument that inter-disciplinary project work, if it is to support significant mathematical learning, must preserve the epistemic integrity of the subject. That is, it should allow the free play and development of mathematical thinking and working, rather than simply apply pre-existing knowledge. Following this, if we are to give interdisciplinary STEM activities greater representation in the mathematics curriculum, we need to investigate how to do this in a coherent way that builds foundational knowledge.

Even where such conditions are not met, the student enthusiasm for STEM tasks, including mathematics tasks, reported by some teachers and students, provides a justification for inclusion of inter-disciplinary STEM activities within the mathematics curriculum. The development of positive dispositions in relation to mathematics, particularly mathematical problem solving (see for example Williams, 2014), is increasingly recognized to be an important curriculum purpose. Further, the argument that these inter-disciplinary activities promote critical and creative thinking beyond what is allowed in traditional approaches, is exemplified to varying extents in these case studies, through the innovative applications of mathematical ideas to situations such as creating strip maps for a race course, investigating the mathematics behind safety ratings, linking billy-cart motion to design features, and developing mathematics to design ramps.

Finally, what comes across in these cases, and through wider experience of these two programs, is the challenge presented in developing a sustainable interdisciplinary STEM program. It seemed there were a number of conditions that needed to be in place to establish such programs, including resources and ongoing professional learning support, champions within the school, and support from leadership within the STEM subjects and in the school generally. The programs were in need of ongoing maintenance, due to the demands of traditional views of curriculum and assessment, timetable constraints, and presumptions about curriculum structures. However, teacher enthusiasm for change was powerful in each case. Historically, integrated curriculum advocacy has never prevailed against disciplinary interests. The question we might ask of current calls for interdisciplinary STEM is: Will it be different this time? Perhaps, with calls for a re-purposing of education, significant changes in the way knowledge is accessed and used, alongside teacher appreciation of student engagement in these activities, inter-disciplinary STEM will become an established phenomenon. Perhaps more importantly, it will provide impetus for reforming school mathematics practices in positive ways.

5.8 Coda

In this chapter we have presented three cases of schools situating mathematics teaching and learning in interdisciplinary STEM settings. The cases illustrate different curriculum arrangements through which students engaged mathematically with authentic design or investigative tasks, different school histories underpinning these, and different challenges faced by mathematics teachers. There are a number of implications we draw out from this comparative study.

5.8.1 The Commonalities in Mathematics Through STEM Despite the Variety of Approaches

The curriculum arrangements differed across the cases in the way the mathematics teachers interacted with technology and science teachers. The key feature however, that we should take as the fundamental principle of STEM-focused mathematics, was in each case the application of mathematics to projects that were 'authentic' and meaningful to students and involved them either in developing mathematics that was new, or applying known mathematics in new ways.

5.8.2 The Role of Disciplines

In no case was it argued that mathematics should evolve into an interdisciplinary practice that was distinct from disciplinary practices. Rather, what was involved was the re-alignment of mathematical thinking and working to more problem oriented and relevant contexts. One might argue that this constitutes a more natural role for disciplinary thinking.

5.8.3 Principles Underpinning Mathematics in Interdisciplinary Settings

From the cases and the literature we argue that for productive mathematics learning in interdisciplinary settings, tasks should (a) engage students' interest, (b) involve problem solving, (c) involve students in using mathematics in unfamiliar and creative ways, and (d) lead to fresh insights into the problem being pursued.

5.8.4 The Challenge for Teachers

Teachers of mathematics have found it challenging to develop productive learning opportunities from STEM problem solving and investigative tasks. This involves a different perspective and perhaps skill set that takes mathematics away from traditional ordered sequencing to a more responsive view of mathematics learning and knowing. However, there was evidence that teachers became more adept at this over time. In Case 2, teachers were able to adapt the tasks to different levels of students' mathematical capabilities. The second challenge, one that teachers seemed to readily adapt to, was the more student centred and open pedagogies often involved in these projects. In Case 3 this adaptation was an explicit focus for the teachers involved.

5.8.5 Conceptual Engagement of Students

There was evidence, in all cases, of students being more enthusiastic about mathematics through these interdisciplinary tasks, and it seemed more conceptually engaged with mathematics. In Case 1 this was aligned with a feeling that the curriculum needed to encourage deeper and more sustained learning, and the ways that contemporary students like to learn. From students' viewpoint, the development of mathematics that was immediately applicable and helpful in problems they felt invested in, provided significant motivation.

5.8.6 The Conditions for Sustainable Innovation

In all cases the development and sustaining of these curriculum innovations depended on high level support from principals, and discipline leaders. In some cases, it involved science, technology and mathematics coordinators who had a similar vision, or in other cases teachers who taught across more than one of these areas.

References

Barnes, M. (2000). 'Magical' moments in mathematics: Insights into the process of coming to know. *For the Learning of Mathematics, 20*(1), 33–43.

Boaler, J. (1997). *Experiencing school mathematics—teaching styles, sex and setting*. Buckingham: Open University Press.

Breiner, J. M., Johnson, C. C., Harkness, S. S., & Koelher, C. M. (2012). What is STEM? A discussion about conceptions of STEM in education and partnerships. *School Science and Mathematics, 112*(1), 3–11.

Bybee, R. W. (2013). *The case for STEM education: Challenges and opportunities*. Arlington, VA: National Science Teachers Association.

Campbell, C., & Jobling, W. (2014). STEM education: Authentic projects which embrace an integrated approach. *Australian Journal of Technology Education, 1*, 29–38.

Clarke, D. (2014). Disciplinary inclusivity in educational research design: Permeability and affordances in STEM education. In *Invited Keynote at the International STEM Conference*, Vancouver, July 2014. http://www.djclarke.iccr.edu.au/.

Dreyfus, T., Hershkowitz, R., & Schwarz, B. (2001). Abstraction in context II: The case of peer interaction. *Cognitive Science Quarterly, 1*(3), 307–368.

Freeman, B., Marginson, S., & Tytler, R. (2015). *The age of STEM: Educational policy and practice across the world in science, technology, engineering and mathematics*. London: Routledge.

Goodrum, D., Hackling, M., & Rennie, L. J. (2001). *The status and quality of teaching and learning of science in Australian schools*. Canberra: Australian Department of Education Science and Training.

Hackling, M., Murcia, K., West, J., & Anderson, K. (2013). *Optimising STEM education in WA schools, Part 1: Summary report*. Edith Cowan Institute for Education. Retrieved from http://www.acola.org.au/PDF/SAF02Consultants/SAF02_STEM_%20FINAL.Pdf.

Honey, M., Pearson, G., & Schweingruber, H. (2014). *STEM integration in K–12 education: Status, prospects, and an agenda for research*. Washington, DC: National Academies Press.

Kieran, C., Guzman, J., Bolleau, A., Tanguay, D., & Drivjers, P. (2008). Orchestrating whole class discussions in algebra with the aid of CAS technology. In O. Figueras, J. Cortina, S. Alatorre, T. Rojano & A. Sepulveda (Eds.), *Proceedings of the 2008 Joint Meeting of the International Group for the Psychology of Mathematics Education and the Group for the Psychology of Mathematics Education Meeting of North America* (vol. 3, pp. 249–256). Mexico: Cinvestat-UMSNH.

Krutetskii, V. (1976). *Psychology of mathematical abilities in schoolchildren*. In J. Kilpatrick, & I. Wirzup (Eds.), (J. Teller, Trans.). Chicago: University of Chicago Press (Original work published in 1968).

Lehrer, R. (2016, October). *Perspectives on elementary STEM education. Keynote address at the forum: Putting STEM education under the microscope*. Melbourne: Deakin University. https://blogs.deakin.edu.au/steme/wp-content/uploads/sites/39/2017/04/Lehrer-Deakin_STEM_2_reduced.pdf.

Lehrer, R. (2017, September). *Considering epistemological junctions when designing for interdisciplinary learning*. Paper presented at the Conference of the European Association for Research in Learning and Instruction (EARLI), Tampere, Finland.

Maaß, K., & Artigue, M. (2013). Implementation of inquiry-based learning in day-to-day teaching: A synthesis. *ZDM Mathematics Education, 45*(6), 779–795.

Marginson, S., Tytler, R., Freeman, B., & Roberts, K. (2013). *STEM: Country comparisons*. Melbourne: The Australian Council of Learned Academies. Retrieved http://www.acola.org.au.

Nardi, E., & Steward, S. (2003). Is mathematics T.I.R.E.D? A profile of quiet dis-affection in secondary mathematics classrooms. *British Educational Research Journal, 29,* 345–366.

National Council. (2015). *National STEM school education strategy, 2016–2026*. Retrieved http://www.educationcouncil.edu.au/site/DefaultSite/filesystem/documents/National%20STEM%20School%20Education%20Strategy.pdf.

Office of the Chief Scientist. (2012). *Health of Australian science*. Canberra: Commonwealth of Australia.

Office of the Chief Scientist. (2014). *Science, technology, engineering and mathematics: Australia's future.* Canberra: Commonwealth of Australia. Retrieved http://www.chiefscientist.gov.au/wpcontent/uploads/STEM_AustraliasFuture_Sept2014_Web.pdf.

Office of the Chief Scientist. (2016). *Australia's STEM workforce: Science, technology, engineering and mathematics*. Canberra: Commonwealth of Australia.

Skemp, R. (1976). Relational understanding and instrumental understanding. *Mathematics Teaching, 77,* 20–26.

Tytler, R. (2007). Reimagining science education: Engaging students in science for Australia's future. *Australian Educational Review, 51*. Melbourne: ACER. Retrieved http://research.acer.edu.au/aer/3/.

Tytler, R., Osborne, J., Williams, G., Tytler, K., & Cripps Clark, J. (2008). *Opening up pathways: Engagement in STEM across the Primary-Secondary school transition*. Canberra: Australian Department of Education, Employment and Workplace Relations. Retrieved http://pandora.nla.gov.au/tep/88047.

Tytler, R., Swanson, D. M., & Appelbaum, P. (2015a). Subject matters of science, technology, engineering, and mathematics. In M. F. He, B. D. Schultz, & W. H. Schubert (Eds.), *The Sage guide to curriculum in education* (pp. 27–35). Thousand Oaks, CA: Sage.

Tytler, R., Symington, D., Williams, G., White, P., Campbell, C., Chittleborough, G. … Dziadkiewicz, N. (2015b). *Building productive partnerships for STEM education: Evaluating the model and outcomes of the scientists and mathematicians in schools program.* Melbourne: Deakin University. Available at: http://www.scientistsinschools.edu.au/downloads/SMiSEvaluationReport2015.pdf.

Vasquez, J. (2015). Beyond the acronym. *Educational Leadership, 72*(4), 10–15.

Venville, G., Wallace, J., Rennie, L. J., & Malone, J. (1998). The integration of science, mathematics, and technology in a discipline-based culture. *School Science and Mathematics, 98*(6), 294–302.

Vygotsky, L. S. (1978). *Mind and society: The development of higher psychological processes.* In M. Cole, V. John-Steiner, S. Scribner, & E. Souberman (Eds.) (J. Teller, Trans.). Cambridge, MA: Harvard University Press.

Williams, G. (2007). Abstracting in the context of spontaneous learning. *Mathematics Education Research Journal (Abstraction, Special Edition), 19*(2), 69–88.

Williams, G. (2002). Associations between mathematically insightful collaborative behaviour and positive affect. In A. Cockburn & E. Nardi (Eds.), *Proceedings of the 26th Annual Conference of the International Group for the Psychology of Mathematics Education* (Vol. 4, pp. 401–408). Norwich: University of East Anglia.

Williams, G. (2014). Optimistic problem-solving activity: Enacting confidence, persistence, and perseverance. *ZDM—The International Journal on Mathematics Education, 46*(3), 407–422. https://doi.org/10.1007/s11858-014-0586-y.

Part II
Focus on Cross-Cutting Skills: A Glass Half-Full?

Pat Drake

Chapter 6
Introduction: A Glass Half Full?

Pat Drake

Abstract The three chapters in this section each exemplify authentic practical problems addressed in learning situations. In so doing they point to the implied questions of scaling up problems so that a wider range of learners and teachers might engage with practical STEM. How can the work of inspirational and creative teachers with high levels of mathematical understanding be extended for wider participation? This part of the book draws out these issues by considering the slipperyness of STEM in a generalised, selective and examination-focused curriculum. In so doing, technology-afforded practice highlights specific areas for teacher development and curriculum liberation. In this section the authors of the chapters are grappling with some difficult issues. The work is all to a greater or lesser extent empirically driven. The studies to which the chapters refer are smaller scale, and although one (Mayes) draws on a project with 20 schools, the other two are set in the context of a single group of students working on a single project.

6.1 Glass Half Full?

Each chapter has a focus on authentic practical problems. LópezLeiva, Pattichis & Celedón-Pattichis present an out-of-school project aimed at Latinx students, in which middle-school students learn to programme digital videos. The other two chapters (Mayes and Sokolowski) both present in-school situations, designed to provoke problem solving that cuts across the Science, Technology, Engineering and Mathematics (STEM) disciplines, and generates different forms of thinking. Sokolowski focuses on developing student mathematical predictions in a teacher stimulated laboratory-based activity, rolling a basketball. Mayes is curriculum-based, and draws on work in the Real STEM Project that identified characteristics of interdisciplinary STEM scenarios to stimulate student-led problem solving. All the authors have written 'glass half full' chapters, enthusiastically reporting on activities that have generated learner engagement, with Sokolowski and Mayes writing about what is possible in school,

P. Drake (✉)
College of Arts and Education, Victoria University, PO Box 14420, Melbourne 8001, Australia
e-mail: Pat.Drake@vu.edu.au

B. Doig et al. (eds.), *Interdisciplinary Mathematics Education*,
ICME-13 Monographs, https://doi.org/10.1007/978-3-030-11066-6_6

and LópezLeiva et al. demonstrating that out of school learning has a part to play in reaching the entire range of learners. Each chapter identifies opportunities for teachers to develop mathematical reasoning with students, although each also notes that students' inclination to mathematical reasoning is infrequent and, or, low level.

The overall challenge for educators, from the chapters in this section, is that interdisciplinary STEM is context dependent and so resists definition. Challenges for schools in providing authentic interdisciplinary curriculum activities in STEM include the high degree of specificity of genuine problems and their connection to specific occupational areas: medicine, engineering and so on. As Hoyles, Noss, Kent, and Bakker (2010) have previously shown, the nature of mathematical thinking at work, is, firstly, technology-related and technology afforded, and second, undertaken in specific ways afforded by the technology and the context. This means that, for example, airline pilots and bankers will engage in mathematical thinking about similar structural problems in entirely different ways. Mathematical reasoning is indeed a process cutting across STEM, but it is undertaken differently in different circumstances.

One way of tackling the difficulty of identifying cross-cutting skills and concepts, is to imagine what professional scientists, engineers, and mathematicians do, and then emulate their activities. This approach leads to exemplar simulations, because what does 'interdisciplinary' mean when there are no bounded disciplines at play in the first place, as tends to be the case in authentic problems? As Engeström (2016, most recently) wisely points out, the whole point of new knowledge is that it is new, it is not known in advance. Nonetheless, each of the papers in this section succeeds in defining STEM in meaningful ways for the purposes of teaching and learning. The papers do this by deciding what characterises thinking in interdisciplinary situations, and then devising activities designed to generate such thinking processes. Each illustrates interdisciplinary STEM activities, and each identifies opportunities for more focused attention on developing reasoning by students.

Another key cross-cutting concept pervading the chapters in this section is the idea of mathematics reaching out to other disciplines. All the authors have addressed the difficulty of pinning mathematical reasoning down—where and when it occurs, what it looks like, and how it can be encouraged. Mayes unpacks, and presents problems, that develop thinking, Sokolowski identifies, through examining student responses to a prediction problem, ways in which students' mathematical reasoning is deficient, pointing to an ideal that he argues should be encouraged. LópezLeiva et al. show how mathematical modelling is integral to the students' design process.

Technology generally cuts across, intersects with, and underpins STEM, and it is indeed technological settings that provide the opportunities for thinking that both Mayes and LópezLeiva et al. consider integral. Sokolowski, however, has offered what seems to start with as a more traditional classroom task, that of predicting the particularities of a rolling basketball, data-gathering using conventional tools of stopwatches and tape measures. This chapter, rather than integrating STEM, attempts to unpick the mathematical reasoning component, and identify students' approaches to mathematical reasoning, with a view to developing ways of teaching this reasoning.

The combination of high level thinking skills with lower level of competence remains, possibly, one of the most significant challenges for organisers of educational experiences, is particularly acute in the case of mathematics, and pervades practice at every level of expertise. We see individual educators addressing this here as best they can, and in stimulating and inspiring ways. But there is an overall challenge in scaling up educational change from individual practices, however exciting, successful, and informative these may be. Innovative practice and thinking shows us what is possible and we can use these ideas for inspiration. What is also necessary is a detailed understanding of contexts for mathematical reasoning, so that organisational curriculum strategy can build contexts in which mathematical reasoning occurs. Sokolowski points out that teachers need this understanding, and suggests the need to encourage professional development for teachers, through engaging them also with investigative experiences.

6.2 Description of the Papers in the Section

Sokolowski describes an activity explicitly designed for high school students in one school to bring together practical, mathematical reasoning, and predictive skills. Scientific methods within interdisciplinary activities were designed to activate students' mathematical reasoning skills. Rather than moving from the observational task of observing a rolling basketball to generating a function to account for the movement of the ball, in this activity students were encouraged to think predictively in the first place before observing the ball in practice. This brought some success in so far as students were generally able to predict some aspects of the motion of the ball, such as its linearity. However, only a small proportion used explicitly mathematical reasoning, i.e. constant rate of change of distance with respect to time. Other responses, although intuitively correct, are more general, and in some cases entirely qualitative, and Sokolowski argues that the activity for students at this level of understanding provides a specific opportunity to focus teaching on mathematising the predictions.

Mayes' contribution draws on a larger scale Real STEM Project in 20 partner middle and high schools, in which beautiful exemplar problems are presented to develop five specific forms of reasoning modality evident in the problem-solving practices of professional scientists, engineers, and mathematicians. The problems, chosen to be 'interdisciplinary' in the sense that each is not bounded by specific disciplines, are explained in detail and aligned with each reasoning modality developed, namely complex systems, scientific model-based, technologic computational, engineering design-based, and mathematical quantitative. Teachers will recognise the passion in Mayes' chapter for the increased engagement and motivation shown by students when tackling 'real' problems. He argues though, that whilst quantitative reasoning has the potential to underpin authentic problem solving, students do not have the skills to adopt it confidently, thus raising the perennial question: which students are 'authentic' problem solving activities actually for? As Dowling (1998) pointed out, so many years ago now, the everyday tends to provide opportunities to practice mundane skills, and the esoteric is for more highly performing students.

LópezLeiva, Pattichis & Celedón-Pattichis take, as their starting point, that schools often do not promote comprehensive understanding of computers and computer programming, and that engineering education is frequently an elective course or simply not offered whatsoever. They argue that this needs to be challenged and present a case of a successful out-of-school technology project in which middle school, predominantly Latinx students, programmed digital videos around self-generated topics. In so-doing, the students engaged in science, technology, mathematics and engineering. Mathematics practices were encouraged through model-eliciting activities, drawn from a framework developed by Lesh, Hoover, Hole, Kelly, and Post (2000) in the early 21st century. The authors' claim is that, through a creative and collaborative process of 'design, model, implement', computational thinking processes emerge through synergies between mathematical modelling, engineering design, computer programming, and student goals.

LópezLeiva et al. point out that technology and engineering practices in schools are not part of the main curriculum. Furthermore, there is a paucity of STEM programs in schools that are inclusive of students from underrepresented groups such as low socio-economic (SES) groups, ethnic minorities, and girls. They suggest that interdisciplinary open-ended problem-solving approaches open to student input are needed to be inclusive of these student populations. An out-of-school project that has recruited largely middle-years Latinx students to learn and practice computer programming adds value to an institutional curriculum provision.

6.3 The Empty Half of the Glass

Implicit in each of these papers is an underlying unease with what STEM means in practice, and discomfort with how practice translates into teaching STEM in schools. Indeed, LópezLeiva et al. argue directly that when schools provide open-ended interdisciplinary activities with student input, students have opportunities to learn challenging mathematics and computer programming and are held to high standards. With the appropriate support to meet these standards and windows to promote student action, underrepresented students can engage at these levels of a demanding curriculum and meet high expectations. This is a two-fold critique, of both curriculum organisation, and participation strategy, each component of which offers a big challenge to STEM policy and design in schools. It is very difficult indeed to organise interdisciplinary STEM activities in educational institutions whose very raison d'être, currently, is the achievement of pre-determined and specified outcomes. But this is not to say that it is not worth doing: the imagined lives of practitioners in STEM, can only be enhanced by more detailed understanding of what they actually do, at all levels, in different cultural settings with contemporary problems. People with STEM qualifications work in a whole range of occupations; and conversely, a whole range of non-STEM occupations require some STEM skills, particularly in

ICT, mathematics, and data handling. As the current Chief Scientist of Australia, Dr. Alan Finkel, recently remarked "As time moves on it becomes increasingly difficult to decide who is, and isn't, a 'STEM worker'" (Office of the Chief Scientist, 2016).

Each chapter has focused on what is identifiable from practice, in practice, and through engagement with social practices, in which mathematical reasoning elides with what might be called STEM problems. The chapter by Mayes goes some way to addressing the challenge through the identification of systems in which STEM features, advocating ways of thinking in these systems. However, as Mayes asserts almost as an aside, students' mathematical skills are less well developed than their scientific thinking skills. From this, arises the other significant educational challenge, that of providing opportunities for engaging in STEM, that are inclusive and enable participation whatever the level of mathematical attainment. This challenge is partially addressed by LópezLeiva et al. who describe their project activities as specifically motivated by the need for inclusivity to draw in younger students' input within high-level curricular tasks. Their data show students can engage in higher levels of the curriculum when provided the opportunity to do so. Their account of the out-of-school project is inspiring, as it suggests ways that informal educational settings (such as those provided via youth clubs) can contribute to educational culture building, enable participants to devise new identities, and expand student self-development expectation.

As asserted earlier, inspiring projects in themselves are not enough to enable curriculum organisers to scale up authentic practice. Is the glass half empty? Well if it is, to identify cross-cutting skills we need to look at the empty half, what is not so far examined here, as well as the full half. The chapters each conclude with a summary of what researchers, practitioners, or policy-makers, should attend to in their future work as a result. LópezLeiva et al. argue for a transformative pedagogy that capitalises on who students are and what they know, mixing with new goal-oriented, open-ended challenges and experiences that enable access to new knowledge, particularly on computer programming and coding. Mayes identifies a lack of STEM expertise in schools, and points to new ways of collaborative working, and ways of extending knowledge, within, and across, schools, specifically identifying the need for explicit opportunities for mathematical modelling. Sokolowski is keen to identify, and focus on, known mathematical weaknesses of students, in particular algebraic thinking. These are all cross-cutting, but they are not skills, for as system needs they go far beyond that. More broadly STEM development in educational settings would include relations between teachers and teachers, learners and learners, and teachers and learners. It would include more emphasis on mathematics for social justice that is studying connections between mathematical reasoning and social division. It must not remain the case that STEM in school serves to minoritise students. It must not remain the case that an absence of confidently qualified teachers means that student access to STEM is of variable quality, even within a single institution. Less qualified teachers need support and training in mathematics if their subject knowledge is weak. Those bringing up-to-date industrial ideas and application need pedagogical support. The cases described in this section each bring both good examples and well-grounded ways of thinking about thinking. It is time now to support a wider network in achieving this.

6.4 Afterword

In editing this section I have been struck by two things. The enthusiasm of the paper authors and their willingness to contribute to this volume as it has taken shape over time has been hugely impressive. Partly this is, I feel, because we all believe in the importance of this volume in making a coherent contribution to the field of interdisciplinary mathematics education. The second striking thing is that there is almost no overlap between the references cited by each author, for although some work on mathematical modelling by Lesh and colleagues over the ten years 1997–2007 has been built upon by both Sokolowski and LópezLeiva et al. clearly there are currently no seminal texts, no shared body of work on which to build our understanding. Thus, I would emphasise the need to develop a research agenda in interdisciplinary mathematics education that begins with a systematic and international review of the field.

References

Dowling, P. (1998). *The sociology of mathematics education: Mathematical myths/Pedagogic Texts (Studies in Mathematics Education Series: 7)*. London: Falmer Press.

Engeström, Y. (2016). *Studies in expansive learning: Learning what is not yet there*. New York: Cambridge University Press.

Hoyles, C., Noss, R., Kent, P., & Bakker, A. (2010). *Improving mathematics at work: The need for techno-mathematical literacies*. London and New York: Routledge.

Lesh, R., Hoover, M., Hole, B., Kelly, A., & Post, T. (2000). Principles for developing thought revealing activities for students and teachers. In A. E. Kelly & R. A. Lesh (Eds.), *Handbook of research design in mathematics and science education* (pp. 591–645). Mahwah, NJ: Lawrence Erlbaum.

Office of the Chief Scientist. (2016). *Australia's STEM Workforce: Science, Technology, Engineering and Mathematic*. Canberra: Australian Government.

Chapter 7
Developing Mathematical Reasoning Using a STEM Platform

Andrzej Sokolowski

Abstract An interdisciplinary laboratory activity involved modelling and interpreting the motion of a rolling ball through the lens of algebraic representation. It was conducted with a group (N = 24) of high school mathematics students. The participants used scientific methods to formulate an algebraic representation of a position for a rolling object on a horizontal surface. While traditional mathematical modelling activities are usually driven by provided data, the technique applied in this study is driven by the phenomenon itself, which serves as a means to verify if the derived algebraic function adheres to the observed behaviour. The results of the study showed that including scientific methods in mathematics interdisciplinary activities may serve as a means to activate, and stimulate, students' reasoning skills, and thus help them integrate the concepts of science and mathematics into a single coherent inquiry. While the study revealed benefits of using hypotheses in interdisciplinary activities, it also opened possibilities of utilizing interdisciplinary laboratories to improve students' mathematical thinking. Suggestions for instructional strategies, as well as suggestions for mathematics curriculum policy makers, are discussed.

Keywords Mathematical reasoning · Scientific inquiry · Modelling · STEM · Hypothesis

7.1 Introduction

Honey, Pearson, and Schweingruber (2014, p. 31) contend that, "learning science entails learning to express the behaviour of natural systems as mathematical models, making this form of integration not merely supportive of but indispensable to learning science." However, the knowledge about a use of mathematics in science education is still fragmentary, and consensus on how to increase the rôle of mathematics in science, especially in physics, is not reached (Uhden, Karam, Pietrocola, & Pospiech, 2012). This study is an attempt to generate a learning experience that would exemplify the

A. Sokolowski (✉)
Lone Star College, Houston, USA
e-mail: andrzej.sokolowski@lonestar.edu

B. Doig et al. (eds.), *Interdisciplinary Mathematics Education*,
ICME-13 Monographs, https://doi.org/10.1007/978-3-030-11066-6_7

rôle of mathematics in science, and simultaneously provide a means of developing students' mathematical reasoning skills. More specifically, the objective of this study is to justify students' mathematical reasoning skills based on how they justified their selection of an algebraic function to describe the position of a rolling basketball.

Mathematical reasoning is characterised by activities such as looking for, and exploring, patterns to understand mathematical structures, and using available resources to solve problems (Schoenfeld, 1992). Mathematical reasoning, merged with scientific conduct possesses the capacity of advancing students' inquiry skills beyond memorisation of facts and procedures, and lead the learners to creating new knowledge (Sokolowski, 2018a). Hypothesis constitutes one of the initial stages of the scientific conduct. Although stating a hypothesis proposes an explanation, the way the investigator formulates and supports the hypothesis, can vary depending on his, or her, mathematical and scientific background. It is considered that the content of a hypothesis can serve as an instrument for justifying the level of the learners' mathematical and scientific reasoning skills and generate suggestions for its advancement.

Observation, experiment, discovery, and conjecture, are as much a part of the practice of teaching and learning mathematics as of any natural science (National Council of Teachers of Mathematics [NCTM], 2000). Several studies (e.g., Berlin & Lee, 2005) have shown that integrated curricula provide opportunities for more relevant learning experiences than traditional teaching methods. Interdisciplinary mathematical activities, that offer contexts during which students apply theorems in practice through experimentation, observation, and conjecture, present an excellent platform for following this recommendation. Through interdisciplinary activities, for example, modelling the periodic motion of an object attached to the spring, or using the motion of two carts, to investigate properties of a system of equations, students engage in observation, discovery, and algebraic model formulation by identifying patterns and using established criteria. Consequently, such experiences can lead to increasing motivation to learn both mathematics and science, and generate positive learner attitudes. Despite a wide diversity of learning opportunities provided by interdisciplinary education, the potential for developing students' mathematical reasoning skills, using interdisciplinary activities, is under-represented. This study attempted to determine whether using scientific methods, and more specifically, scientific hypotheses, in an interdisciplinary mathematics activity, can challenge students' mathematical reasoning.

Interdisciplinary mathematics education encompasses multidimensional types of integrated learning. One of the types of interdisciplinary activities is a block of Science, Technology, Engineering, and Mathematics (STEM). In this study, students will experience merging the attributes of a linear function with the properties of motion with a constant velocity. The activity will be conducted in a mathematics class. While STEM can merge several disciplines, research shows that mathematics is not fully exercised in that paradigm (English & King, 2015; Tytler, Prain, & Peterson, 2007), therefore attempts to increase its contribution are made. Since the primary type of inquiry in STEM activities gravitates toward scientific methods, the question that arose was whether inducing these methods in mathematical STEM activities,

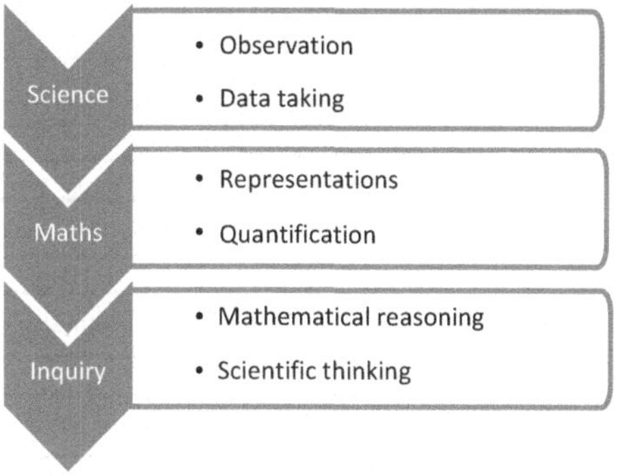

Fig. 7.1 Proposed pathway to develop students' inquiry skills

could serve as a catalyst for fostering students' mathematical reasoning. Although, at first, the idea seemed to propagate further, the scientific methods in mathematics, rather than mathematical reasoning in STEM, experimenting with such a pragmatic framework appeared to be a promising endeavour benefiting, not only students' mathematical thinking, but also helping them develop scientific inquiry skills, and a general disposition to undertake more complex STEM projects. The idea of merging the tools of mathematics with science, during this activity, is schematically illustrated in Fig. 7.1.

While the details of the pathway are further developed in Fig. 7.2, the main difference between what traditional mathematical modelling is, and what the proposed method offers, is the inductive character of the inquiry, and a constant intertwining between the contents under investigation. Application of mathematics is not driven by the provided data, but by observation and by a mapping of the behaviour of the physical quantities with the best fit to an algebraic function. This study will focus on investigating how students merge mathematics with science while formulating hypotheses for their investigations.

7.2 Theoretical Framework of the Activity Design

The activity was formulated using an integrated modelling scheme (see Fig. 7.2), that the author had previously developed, by meta-analysis of findings of the learning effects of interdisciplinary modelling activities (Sokolowski, 2015). It was argued that mathematics and science concepts need to be explicitly flagged, and linked, during the activities to enable a deeper analysis of their relationships and to strengthen their mutual interpretation. While the final product of the investigation is an algebraic function, the knowledge that students are to gain from the investigation is of a dual nature; the enacted algebraic representation is to serve as a means to learn more about the scientific nature of the experiment.

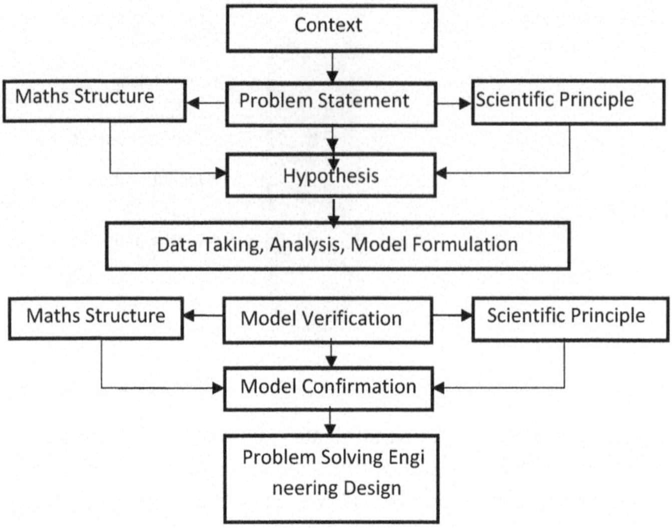

Fig. 7.2 Integrated STEM modelling cycle (Sokolowski, 2015)

Mathematical activities can be structured in various ways, typically supported by mathematical modelling cycles. Analysis of several such cycles, (see, e.g. Greefrath & Vorhölter, 2016; Maaß, 2006) revealed that these cycles often lead students to finding a unique solution to a given problem, rather than providing a means for developing their scientific and mathematical thinking. Since the idea of using scientific methods has begun to gain more attention in STEM education, thus also in STEM activities conducted in mathematics classes, several theoretical frameworks emerged. For example, Kelley and Knowles (2016) designed an integrated STEM framework that included scientific inquiry as one of its main pillars, Kennedy and Odell (2014) suggested that high-quality STEM education programmes should promote scientific inquiry, that includes both rigorous mathematics and science instruction. The literature offers more examples of schematic diagrams, yet detailed practical examples are rarely found.

While providing students with opportunities to use science contexts as a platform to apply the tools of mathematics is not a new idea, and literature provides resources on organizing such activities (e.g., Berlin & Lee, op cit), using a hypothesis, along with the goal of inducing it in interdisciplinary mathematical activities to develop mathematical reasoning, is rare (Sokolowski, op cit). Several researchers (e.g., Crouch & Haines, 2004; Diefes-Dux, Zawojewski, Hjalmarson, & Cardella, 2012) pointed out that students' skills in formulating hypotheses, organising the proof, and validating elicited mathematical structures, are weak, but they did not suggest, explicitly, how to improve these skills. Formulating a hypothesis as a prediction of how a system's outputs depend on the inputs, is linked to identifying independent and dependent variables. Research on modelling in mathematics (Carrejo & Marshall, 2007) has reported students' difficulties with identifying variables, and classifying the variables, to formulate mathematical constructs. Inviting students to hypothesise about laboratory outcomes induces the idea of predicting the behaviour of the experiment, and thus, it makes the students view the physical quantities as components of certain algebraic structures (functions) whose behaviour is consistent with that of the experiment in a mathematical sense. Figure 7.2 illustrates the general framework that makes the stages of connecting mathematical and scientific aspects of a real laboratory more explicit.

It is suggested that the problem statement be provided by the instructor, and students' tasks are to extract, and then merge, relevant concepts from both disciplines into one symbolic representation that is consistent with scientific behaviour of the variables of interest, and with further hypothetical analysis of the scientific nature of the experiment, as viewed through the properties of the algebraic form. The framework explicitly highlights these stages where scientific principles, and mathematical structures, are to be extracted from the laboratory context and integrated into a symbolic representation.

The instructional support offered to students' prompts, and self-directs, actions that supported formulation of the symbolic (algebraic) form of the basketball's position. While the scheme shows the flow of tasks in one direction, the students were encouraged to repeat it if the verification process disqualifies the elicited algebraic model. Prompts for revisions were provided in the instructional support.

The main science concept employed in this experiment was the idea of motion with a constant speed. While in physics, the concept of velocity is used to describe rate of change of an object's position, in this laboratory, the concept of speed was applied because the students were not supposed to describe the basketball's direction of motion but its magnitude only. The idea of modelling an object's speed to challenge students' mathematical reasoning, was selected for the following reasons:

(a) Students who took part in the study already possessed a conceptual background for this concept from their physics classes.
(b) Speed (the scalar version of velocity) is also often applied in mathematics textbooks (see Larson, 2005).
(c) The idea of motion with a constant speed carries rather low cognitive load, thus no extra introduction to the mathematical modelling of this type of motion was necessary.

Since in science or physics courses the students used the terms speed and velocity to describe object's rate of change of distance or position, the instructor explained the differences to assure a cohesive terminology between both disciplines.

By merging the knowledge of different disciplines, STEM exploratory activity usually offers students an opportunity of deriving new knowledge (Sokolowski, 2018b).What was the anticipated new knowledge generated during this activity? In physics classes, students focus on investigating an object's constant rate of change of position over time, as evidence of uniform motion (e.g. Tipler & Mosca, 2007). However, in this interdisciplinary mathematics activity, in addition to hypothesising the form of the function based on the behaviour of the variables, the students constructed an algebraic function supported by observation, data taking, and graph

Fig. 7.3 Epistemology of the lab activity

sketching. Students merged both streams of knowledge into one, coherent, representation, as depicted in Fig. 7.3.

Students were to formulate hypotheses for the stated problem, and also to reflect on the hypotheses by refuting or accepting them, considering the adherence of the data to the algebraic forms. Thus, while in physics, students investigate the properties of motion, and represent the properties with formulae, this activity represented the motion with algebraic functions, that provided more opportunities for extending the analysis of the motion beyond the physical domain of the experiment. As an example of such an extension, the students were required to use the functions to find either the time, or the position, of the rolling basketball. They were also to apply function transformations to construct new position functions based on pre-arranged new conditions. The purpose of this inclusion was to increase the interdisciplinary character of the investigation. More specifically:

(a) Have the students build a tangible image of the experimental design as a basis for constructing new representations that would eventually help them with problem-solving.
(b) Provide applications of function transformations in non-traditional context.

A short discussion is needed with the students about assuring that the motion of the basketball will be close to uniform. The ball will slow down due to air resistance and other factors, like the floor not being of a uniform texture. Because the ball will roll, rolling friction is needed and it helps to roll. To assure a uniform speed, the students might be asked to push the ball so that it moves 20 m but take the time measurements over first 10 m of its motion.

7.2.1 Conduct of the Laboratory

Before the laboratory activity, the students reviewed the basic properties of linear functions, along with sketching, and finding the constant rate, when the values of quantities were given. Due to time constraints, the properties of motion with a constant velocity, as seen from the physics point of view, were not explicitly reviewed, nor were they taught.

The teacher opened the lesson by informing the students that they would observe the motion of a rolling basketball, and then hypothesise the form of the function equation, record data (i.e., measure the time the rolling basketball passed certain points), and formulate the function equation. The teacher demonstrated the motion in class, then, he handed out instructional supports to each student. After observing the motion, but still in the classroom, the students were invited to formulate their hypotheses. This arrangement was to assure that the students' hypotheses were individually formulated, thus, the data validity was secured. After the students finished formulating their hypotheses, the teacher divided the class into four groups of six students per group. Each group received four stopwatches, a tape measure, and a basketball. The teacher explained that each group must create a set of at least four co-ordinates that would depict the motion of the basketball with time and position as the variables. The distance of the motion of the ball was 10 m. Thus, some groups worked in the hallway. After plotting the points on a position-time axes, the students were asked to select a function, based on the data distribution, and then compute the coefficient of determination and correlation coefficient for the selected functions to justify, statistically, the choice of function. The teacher emphasised that although the data gathered by the groups would be the same, the best-fit curves, as sketched by hand, and their equations, might vary. The teacher briefly reviewed the technique of sketching the line of best fit. Once the students started organizing the process of data collection, the teacher took on the role of facilitator.

7.2.2 Methods

This study can be classified as one-group quasi-experimental (Shadish, Cook, & Campbell, 2002). Randomisation of participants was not possible due to the low school population in which the study was conducted. A quasi-experimental study shares many similarities with experimental design.

The study attempted to address the following research question: Can a verbal formulation of a hypothesis, and its verification, serve as a means of developing students' mathematical reasoning?

The participants in the study consisted of a group of 24 mathematics students (11 males and 13 females, age range 16–17 years) from a suburban high school. These students did not have a formal prior experience with modelling activities in their mathematics classes, nor with writing hypotheses. Five of these students (21%) were concurrently taking a physics course, and they had previously studied different types of motion in their physics class. The evaluation instrument consisted of an analysis of students' verbal formulation of hypotheses. The quality of the verbal responses served as an indicator of their quality of mathematical reasoning. The students then observed the experiment, collected data, and formulated algebraic representations that described the process of the experiment. They sketched the best-fit line by hand, and formulated the equation of the function. They also used technology to find a respective regression line. They computed the coefficient of determination for the best-fit line, reflected on the precision and accuracy of the data, and their hypotheses. Furthermore, they attempted to identify the sources of incorrect prior thinking in cases where their justifications were not supported by the data and the derived models.

7.3 Data Analysis

Students were expected to provide a hypothesis for the following problem: What type of function can be used to model the position of a basketball rolling along a floor if air resistance is ignored? Support your answer using your mathematics and science background.

All students selected a linear function to model the position of the rolling basketball. The depth of conceptual support, and the extent to which the students used mathematical and scientific terminology and reasoning, showed a range of diversity of reasoning across the students. Since the purpose of the study was to find out whether formulating hypotheses could be used to justify students' mathematical reasoning, the quality of the responses was not quantified but was clustered in two groups to reflect their common features. The students were asked to support their hypotheses using explicitly scientific and mathematical terminology, but their responses blended the terminology, of both of these disciplines, in verbal structures. The question that arose during the analyses of the responses was how to differentiate between students' scientific and mathematical reasoning? Is a different terminology used (scientific versus mathematical) sufficient? If so, would using, for example, constant rate, constitute mathematical or scientific reasoning? The literature about justifying the difference between mathematical and scientific reasoning is limited. As a reference, I have used schemes developed by national review physics committees. Viewing these recommendations, it was decided that making these distinctions was not necessary, because, for example, the term constant rate is used in both mathematics and science. Thus, to complete the preliminary analysis, separate categories for these responses, were not used, but rather, the quality and depth of the individual response were considered as a factor for reasoning evaluation. This analysis resulted in formulating two response groups. Group 1 constituted responses with more accurate properties of linear functions, such as, linear, or rate, or slope, and sufficient scientific support, like constant speed, and constant rate of change of position (these are summarised verbatim in Table 7.1) and Group 2 (see Table 7.2) included samples of more general responses, sometimes stating an overall dependence of the variables of interest.

7.3.1 Descriptive Analysis

Among the 24 responses, six students (25%) showed that they explicitly considered the properties of linear functions along with the principle of motion with a constant speed to support the function selection. These students (e.g., Student 2, 3, and 4 listed in Table 7.1) merged the attributes of linear functions effectively with patterns of observable motion, thereby making the justifications precise. They extracted the independent and dependent variables from the experiment, and embedded the variables in the algebraic rule formulation. The responses listed in Table 7.2

Table 7.1 Responses that included properties of linear functions

Student	Response
1	If the ball is rolled on the floor, there would be a linear equation that has a value of b of zero with a positive slope
2	The function shown by basketball rolling on the floor should be linear because for every unit of time, the distance the basketball rolls increases by the same magnitude. Also, there is no horizontal acceleration, so the ball should be moving at a constant speed if the surface is frictionless
3	The function to represent the data would be linear because the position and time will increase at a constant rate
4	The ideal function to model the basketball position would be a linear function. If the force of friction is neglected, the ball would roll at a constant rate, making a linear function the best way to connect the points
5	A linear function because the rate of change will be constant
6	The graph will be linear due to the rate

Table 7.2 Responses that did not include properties of linear functions

Student	Response
1	As the ball rolled along the floor, the ball would have a strong positive correlation because as each second passed the metres increased, until someone caught the ball
2	Just because the ball will eventually slow down, then that does not mean that a regression line will be a square root function or quadratic function
3	The line on the graph would be most likely linear because the amount of time it takes the ball to roll further would always go up
4	The function should be linear because the distance the basketball moves increases by the same magnitude
5	If the basketball runs across the floor, ignoring air resistance without stopping, the data shall compute only a linear function
6	The ball should plateau at some point because gravity forces the ball to stop rolling
7	If basketball is rolled at a constant speed, ignoring air resistance, then the data will produce a linear graph
8	It will be linear because as the time increases so would the distance unless there is a significant amount of friction
9	As the ball rolls, there are forces that act upon it causing it to gradually reduce its speed to a complete stop. Therefore, the equation would be a linear
10	If the ball increases in the distance, then time increases

Table 7.3 Percentages of students using function attributes

Algebraic terms	Number of students and percent of the sample
Linear	18 (75%)
Constant	6 (25%)
Rate	5 (21%)
Slope	2 (8%)
Function	11 (44%)

show that the main attribute of linear functions—that is, a constant inclination (gradient) representing a constant speed or constant rate of change of distance over time—was not used very often by the students. This weakness might imply insufficient conceptualisation of the slope idea when it was first introduced to the students. While some students extracted the concept of a linear relation, between the ball's position and time, see for example Student 3 (Table 7.2), they did not convert these statements into a more precise mathematical representation. The remaining part of the group (N = 18, 75%) supported their claims even more loosely.

Five students from this group (28%), suggested using an increasing function due to an increase in the basketball's position which was correct. Their supporting statements though, lacked the specific terms that would warrant the application of the linear function. Thus, their claims were not evaluated, as satisfactory. Motion attributes, such as a constant rate, represented by a constant slope, were also not visible among this group of responses. The support statements were vague, showing these students' weaknesses in understanding the conceptual merit, and attributes, of linear functions. Table 7.3 summarizes the percentage of responses using the specific properties of linear functions for the entire group of students (N = 24).

The students were also encouraged to compute the coefficient of determination, for the data distribution, to support their function selection, and to justify their hypotheses. The coefficient computed by using technology was high and in the range of 80–92% for all groups.

The number of students, who conceptualised their selected function using sound mathematical terminology, ranged from 8% (N = 2) who used the term slope, to 75% (N = 18) who used the term linear. Perhaps the percentage would be higher if these students were instructed on what function attributes to look for, or how to use these attributes, to verbalize such justifications.

7.3.2 *Inferential Analysis*

One of the themes that emerged from the study was the effect of the interdisciplinary background—in this laboratory about kinematics—on supporting students' mathematical reasoning. It was anticipated that to mark the activity as a productive learning experience, no scientific introduction of the idea of uniform motion was necessary. The laboratory results were not consistent with this assumption. While all students formulated the position function for the rolling basketball correctly, the way that they verbalised the scientific part of their responses, revealed a lack of sufficient scientific terminology and understanding of different types of motion. For example, Student 6 (Table 7.2) said, "The ball should plateau at some point because gravity forces the ball to stop rolling," which is not entirely true, because it is the strength of resistance (e.g. the air resistance) that stops the basketball. While the force of friction is dependent on the force of gravity through the normal force exerted on the moving ball, the direction of the force of gravity was perpendicular to the ball's velocity and therefore it had no direct bearing on the ball's speed.

Hypotheses reflect closely on the problem under investigation. Thus, while formulating hypothesis in an interdisciplinary setting, students are expected to recall prior knowledge and provide a viable answer coherent with mathematical constraints and the natural behaviour of the system under investigation. Asking students to state hypothesis should help them to focus on a narrower area of inquiry. Since interdisciplinary mathematics activities can be dynamic, students can also be required to classify the quantities as dependent or independent. The integrated modelling scheme (see Fig. 7.1) suggested that students apply mathematical reasoning to describe the behaviour of the phenomena and support the description by using scientific reasoning. The task of merging both disciplines while formulating a hypothesis was explicitly addressed in the instructional support provided to the students.

The author believes that providing mathematics students with diverse cases to extract correct forms of algebraic functions, or their attributes, or theorems, in interdisciplinary activities, can serve as a means of improving their mathematical reasoning skills. The challenge is the extent to which similar tasks can be designed to assist teachers to understand the development of mathematical reasoning using interdisciplinary learning, and how to extract the prompts of high-quality reasoning skills from students' work. The question of interest can also be the balance between, purely abstract mathematical understandings, and, the ability to apply the understanding in contexts. For instance, how to quantify students' understanding of the Mean Value Theorem using the idea of a pure algebraic function versus using, for example, a velocity function and subsequent acceleration function? To what extent can mathematical reasoning be taught and be developed without a context? Alternatively, how do we measure the learning effects when pure algebraic representations are given, as opposed to using context to uncover the representations, and then analyse them?

7.4 In Search of Improving the Learning Experience

Analysing students' responses, we have learned that reviewing, or even introducing, scientific background would benefit students' learning experiences, and perhaps more efficiently help them integrate the knowledge of both disciplines. Therefore, discussing the contexts in more detail, prior to the laboratory is recommended. This idea, is further supported, by considering the manner in which students, who were concurrently taking a physics course, formulated their hypotheses (there were five such students in the group, 21%). These students' responses (see Table 7.1, Student 2, 3, 4, 5, and Table 7.2, Student 8) displayed a higher level of understanding of motion, and their justifications for using linear functions to model uniform motion, were more consistent with the conditions of the motion. One could conclude that all students' reasoning skills would be deeper if a relevant review about properties of motion were provided. While the assumption that all students would realize that a constant speed calls for a linear function to be applied, the laboratory showed that not all students had that understanding. The laboratory also contained a section where the students reflected on their hypotheses, and provided possible sources of error.

The scientific method, by which scientists endeavour to construct an accurate representation of the world, consists of the following four elements: observation, formulation of a hypothesis that proposes an explanation of the phenomenon, the performance of an experiment, and the testing of the hypothesis. During the investigation, the rôle of a hypothesis is to confirm, or correct, an investigator's understanding of what is the content of the experiment. A hypothesis can also be perceived as a provisional idea that requires evaluation. For a valid assessment, it needs to be defined in operational terms referring to the specifics of the investigation, and it usually requires a formal scientific experiment designed and organised by the investigator. A confirmed hypothesis has the potential to become a law, and it is often expressed in a mathematical form (Simon, 2012). Moreover, a hypothesis can be perceived as the investigator's proposed theory, explaining why something happens, based on the researcher's prior knowledge (Felder & Brent, 2004). Thus, to increase the prestige of students' work, they can be asked to formulate a theory based on their discoveries. This task would serve as a factor strengthening their confidence, and encourage their own investigations. Formulating and proving, or disproving, a hypothesis might take various forms; for example, it can involve a statistical procedure, as developed by Fisher (1955), in which a null hypothesis is formulated and tested statistically to be rejected, or accepted, depending on whether results fall within an established confidence interval. Since, using statistical methods requires advanced analytical apparatus, this might not be accessible to all high school mathematics students, thus, these methods were not employed, but instead, students formulated their hypotheses verbally, using their mathematical and scientific background. A hypothesis can also take the form of existential statements claiming that some instances of the phenomenon under investigation have some characteristics and causal explanation (Popper, 2005). As an existential statement, a hypothesis can be formulated verbally and proved, or disproved, following the scientific approach, and this form was employed in this laboratory.

The action of formulating a hypothesis is closely related to developing a prediction. While a hypothesis proposes an explanation for some puzzling observation, a prediction is defined as an expected quantitative outcome of a test of some elements of the hypothesis (Lawson, Oehrtman, & Jensen, 2008). Being of a quantitative manner, prediction can also be included in mathematical activities to support a hypothesis, and justify, for example, students' estimation skills. In this laboratory, the students were asked to predict the magnitude of the speed in metres per second based on the observable motion. The range of their predictions varied more than expected, but this was accounted for by the fact that in everyday experience, these students use feet per second rather than metres per second to describe the rate of change of distance.

7.5 Discussion

The purpose of the laboratory was to have students experience the process of merging the content of mathematics and science in an interdisciplinary setting and generate new knowledge merging both disciplines. From a scientific point of view, students experienced that, motion with a constant speed, implies a constant slope. From a mathematics perspective, they learned that a constant rate of change, of quantities of interest, warrants application of linear functions.

This merging was to help students view the content of mathematics, not as a static subject bounded by procedures and facts, but as a dynamic process of creating new knowledge grounded in natural settings. Provision of instructional support guided the students through the merging stage, as well as supplying them with opportunities to reflect on their thought processes, while leaving room for their input. The idea of conducting the interdisciplinary activity was intended to assist students to link both disciplines, and the task of formulating hypotheses, was meant to provide prompts for evaluating the students' mathematical reasoning skills. Evidently, having students state a hypothesis, and reflect on it, served as a means of activating their thinking. While the students formulated their hypotheses before working on the activity, there is still an unanswered question about how the experience augmented their reasoning skills, in particular of those students whose responses were listed in Table 7.2. The question of how to best formulate the problem statement, or the prompts, during the activity, so that students' engagement and learning is maximized is not completely answered. Did observing the experiment, collecting the data, formulating the algebraic representation, and proving, or disproving, the hypothesis, deepen the integration of knowledge? Did it help to emphasise the importance of the mathematical aspect, or did it emphasise a discovery of motion with a constant speed? While either of the goals would be satisfactory, evaluating these effects would help with designing other interdisciplinary laboratories. The preliminary results of the study seem to be encouraging, but more such activities need to be designed and conducted to better reflect on the learners' cognitive processes.

There are several general recommendations that emerged from this study. First, there seems to be a need for developing assessment instruments that would evaluate students' skills of merging the concepts of mathematics with extracted quantities from interdisciplinary settings, while simultaneously promoting their creativity and reasoning. It is apparent that evaluating students' procedural skills does not provide the desired evidence. All of the students formulated the algebraic function correctly, however, based on their prior experiences, reasoning skills of only six of these students were rated as satisfactory. Being successful on solving problems that often require finding a unique solution, does not necessarily prove adequate reasoning skills, but simply just proves that the student has adequate algebraic skills. Typical problem solving, which often requires memorisation of procedures, does not seem to support reasoning in a sense that students would use in their science classes. Designing mathematical assessment items that require the subtle area of merging mathematics with interdisciplinary, physical quantities, is a task worthy of further development.

The laboratory activity also revealed that, to encourage mathematical reasoning, there should be more emphasis on using function properties, or attributes, to select the most appropriate representation for given data. Statistical tests are necessary, but they do not provide a bridge between observable, or measurable, variables, and their graphical depiction, and the task of using statistical tests, often performed by technology, do not nurture the bridging. There should be review sections in mathematics textbooks designated to contrast various algebraic concepts, and challenge their applicability, and their limitations in different real situations. Sketching functions and finding their equations without context perfects procedural skills, but context-driven scenarios move the learning to a more sophisticated level. In such settings, the procedural skills will serve as a prelude to a deeper conceptual understanding of mathematical concepts, rather than an absolute means of generating the final product.

The element of hypothesizing, in this activity, had a dual purpose: it was intended to lead the students' thinking processes and serve as an instrument to justify their mathematical reasoning, and thus provide prompts for laboratory improvements. The effects of developing a hypothesis, along with conducting a scientific process, do not exhaust the learning objectives that an interdisciplinary laboratory can target. Simply stating that a hypothesis is correct, or incorrect, limits students' input and their share in the conduct of the laboratory. An interdisciplinary laboratory provides ample opportunity for students to reflect more deeply on their prior learning experiences, and correct, or modify, these experiences, if needed. How to accomplish this goal? After concluding the laboratory, students should be provided with problem-solving questions that would link the acquired experience with textbook problems, or other assessments, not necessarily referring to mathematics textbooks.

Finally, the data analysis revealed that more could be done in highlighting the interdisciplinary principle under investigation prior to the laboratory. Thus, supplying students with more tools, that they could have at their disposal, for the laboratory, would improve the laboratory outcomes, and would very likely be appreciated by the learners. For example, the categorisation of variables, as dependent or independent, is often related to plotting functions in the Cartesian plane, and such categorisation is explicitly given to students in their mathematics classes. Such a level of categorisation is not sufficient for successful integration of mathematics with other disciplines. It is seen that students' deeper understanding of the principle under investigation is necessary, to make the distinctions in real contexts. An interdisciplinary relationship provides the added benefit of improving students' general learning experiences, beyond the sole domain of mathematics. This feature however, cannot be easily implemented in mathematics daily routines because the curricula, scope, and sequences, are strictly designed. Perhaps, interdisciplinary activities should be conducted during an independently formulated subject, called for example mathematical reasoning, where more time could be devoted to developing the contextual background and intertwining the concepts. Such courses, organised in parallel with mathematics, appear promising, and worthy of further consideration.

While hypotheses in mathematical interdisciplinary activities will most likely be verbalised with the aim of testing mathematical structures, and reflecting on how well the structures describe a system's behaviour, there is a need for a more detailed discussion on how to formulate a hypothesis in interdisciplinary mathematics activities. The final products of activities in mathematics classes can take various forms, ranging from formulating abstract mathematical representations to building artefacts. In the proposed integrated modelling cycle (see Fig. 7.1), it is assumed that the final product will be a mathematical representation, often in the form of an algebraic function. In this study, the stance was that the way the learner supports the hypothesis, along with using function attributes, and merging them with the system behaviour, can be used to justify students' mathematical reasoning as it pertains to understanding applications of linear functions to motion problems. Hypothesising in mathematics requires a conceptual understanding of functions, or ratios and proportions, domain, range, formulation of maximum, limits, intercepts, and so forth. For instance, if students are to formulate a function representing Newton's second law of motion, then in mathematics classes, their hypotheses will attempt to answer questions about the type of algebraic function that can be used to describe the mathematical relation between an object's acceleration and the net force. Students might predict a linear relation, because if the net force increases, so will the object's acceleration. A plotted graph will resemble a linear function, and its slope will show as the proportionality constant or as the object's mass. How will this process differ from what students learn in their science classes? In science classes, students hypothesise the type of proportionality that can be formulated due to investigating change of object's acceleration when the net force acting on it changes (Tipler, op cit). The term function, as studied in mathematics classes, is rarely used in science, and the scientific hypotheses will merely require the students to find if the relation is directly or indirectly proportional. Formulations of these hypotheses are mutually inclusive, and one can be perceived as a complement to the other, yet their natures are subject-domain dependent, and they appear to take a narrow meaning when they are discipline bounded. A proposed method of integrating the contents is to have students merge the hypotheses into one statement, consistent with mathematical terminology and rules, and with scientific laws.

It seems that merging both disciplines in a unified learning experience, and projecting the phenomenon analysis, using the methods of all subjects involved simultaneously, is a way to knowledge exploration and acquisition.

7.5.1 Suggestions for Further Research

The students contended that their hypotheses were correct, focusing on predicting linear graphs for the motion. They showed a high level of inaccuracy in time recording (the values of the coefficient of determination computed using a graphing calculator ranged from 80 to 92%) that affected the speed computations. They assumed that stating the statistical value of the error was sufficient, and therefore they did not attempt to correct their thinking, although this was assumed to take place.

While this deficiency can be eliminated, by designing a laboratory when the time could be detected by using a photo-gate, the answer how to correct or improve students' reasoning remains open. Will the students enhance their reasoning by conducting more similar interdisciplinary laboratories, or will they improve the reasoning by having the instructor pinpointing elements of the laboratory analysis and computations that were determinant to enrich their algebraic thinking?

The purpose of the study was to investigate students reasoning based on how they formulated their hypotheses, not how they corrected their thinking, therefore it was considered that the study served its purpose. Yet, the effects of interdisciplinary activities can be extended to include the measure of change, perhaps by interviewing students before, and after, the laboratory, and extracting changes in reasoning. This is certainly an area to pursue further.

The tasks of hypothesizing a possible graph for identified variables in a given experiment, along with providing verbal support, are included in science textbooks (see, for example, Etkina, 2010, p. 54) and they are included on national physics examinations (see AP Physics). According to reviewers, physics students have difficulties with producing such representations. More specifically, they do not only face difficulties with identifying and classifying variables, in a mathematical sense, but also with selecting a correct function, given by a curvature, or plot. These concepts are not frequently exercised in mathematics classes, although they represent core mathematical ideas. Instead, students analyse pure functions without contexts that probably supports learning by rote, not genuine understanding. Such teaching does not link mathematics with real world situations, and does not help students with applying these ideas in other subjects Consequently, such teaching does not help the learners to broaden their view about applications of mathematics, and does not encourage them to study mathematics for understanding. Learners need to be provided with contexts, and be invited to extract the applications of theorems studied, from contexts. Can science communities be invited to initiate a better alignment of the mathematics they use in their science classes with mathematics that students study in their mathematics classes? There is certainly room for a change.

The task of verbalizing attributes of algebraic functions using contexts is silent in mathematical curricula, and it is seen that more attention could be designated to develop this skill. Students need to realize that each type of function has its own rules for sketching, or finding intercepts. More importantly, they need to realize that function attributes (e.g. existence of a horizontal asymptote, or a piece-wise nature) classify the functions to describe different phenomena, and vice versa, that a specific property of the phenomenon's behaviour, will demand a specific algebraic representation. In typical mathematics textbooks, problems on certain kinds of function are usually clustered at the ends of chapters according to the types of function, or ideas, that the functions describe. Perhaps providing a diversity of assessments, where students would extract and pin-point specific attributes; e.g., constant percent rate, a local maximum or constant period of motion, would offer students more room to reason mathematically when context is provided.

The nature of the mathematical tools and systems of representation, available to students, determine the depth and breadth of learning about core ideas in science because mathematical forms correspond to forms of understanding natural systems (Quinn, Schweingruber, & Keller, 2012). It is expected that confirmed algebraic representations, resulting from similar investigations, can be further used to delve deeper into behaviour of scientific systems, thereby triggering opportunities for producing creative thinkers ready to generate new knowledge. This study, and also others from prior research, shows that students need to have a sufficient scientific background to be able to take advantage of the full range of learning benefits, that the method of integrating, offers. While verifying with students if certain concepts were covered, in the respective science class is a suggestion, this might not apply to students who are not concurrently enrolled in that particular science course. What would be the best solution? It seems that creating a separate course that would allow teachers, to devote extra time for bringing up the scientific contexts, would benefit the learners the most.

There are feasible ways of improving the quality of teaching and learning mathematics and science. It is hoped that this study offers suggestions that are worthy of being given further consideration.

References

Berlin, D. F., & Lee, H. (2005). Integrating science and mathematics education: Historical analysis. *School Science and Mathematics, 105*(1), 15–24.

Carrejo, D. J., & Marshall, J. (2007). What is mathematical modelling? Exploring prospective teachers' use of experiments to connect mathematics to the study of motion. *Mathematics Education Research Journal, 19*(1), 45–76.

Crouch, R., & Haines, C. (2004). Mathematical modelling: Transitions between the real world and the mathematical model. *International Journal of Mathematical Education in Science and Technology, 35*(2), 197–206.

Diefes-Dux, H. A., Zawojewski, J. S., Hjalmarson, M. A., & Cardella, M. E. (2012). A framework for analyzing feedback in a formative assessment system for mathematical modeling problems. *Journal of Engineering Education, 101*(2), 375.

English, L. D., & King, D. T. (2015). STEM learning through engineering design: Fourth-grade students' investigations in aerospace. *International Journal of STEM Education, 2*(1), 1.

Etkina, E. (2010). Pedagogical content knowledge and preparation of high school physics teachers. *Physics Review, 6*(2), 020110.

Felder, R. M., & Brent, R. (2004). The intellectual development of science and engineering students. Part 2: Teaching to promote growth. *Journal of Engineering Education 93*(4), 279.

Fisher, R. (1955). Statistical methods and scientific induction. *Journal of the Royal Statistical Society. Series B (Methodological) 17*(1), 69–78.
Greefrath, G., & Vorhölter, K. (2016). *Teaching and learning mathematical modelling: Approaches and developments from German speaking countries*. New York: Springer International.
Honey, M., Pearson, G., & Schweingruber, H. (Eds.). (2014). *STEM integration in K-12 education: Status, prospects, and an agenda for research*. Washington, D.C.: National Academies Press.
Kelley, T. R., & Knowles, J. G. (2016). A conceptual framework for integrated STEM education. *International Journal of STEM Education, 3*(1), 1–11.
Kennedy, T., & Odell, M. (2014). Engaging students in STEM education. *Science Education International, 25*(3), 246–258.
Larson, R. (2005). *PreAlgebra*. Evanston, IL: McDougal Littell.
Lawson, A. E., Oehrtman, M., & Jensen, J. (2008). Connecting science and mathematics: The nature of scientific and statistical hypothesis testing. *International Journal of Science and Mathematics Education, 6*(2), 405–416.
Maaß, K. (2006). What are modelling competencies? *ZDM Mathematics Education, 38*(2), 113–142.
NCTM. (2000). *Principles and standards for school mathematics*. Reston, VA: National Council of Teachers of Mathematics.
Popper, K. (2005). *The logic of scientific discovery*. New York: Routledge.
Quinn, H., Schweingruber, H., & Keller, T. (Eds.). (2012). *A framework for K-12 science education: Practices, crosscutting concepts, and core ideas*. Washington, D.C.: National Academies Press.
Schoenfeld, A. H. (1992). Learning to think mathematically: Problem solving, metacognition, and sense making in mathematics. In D. Grouws (Ed.), *Handbook of research on mathematics teaching and learning* (pp. 334–370). New York: Macmillan.
Shadish, W. R., Cook, T. D., & Campbell, D. T. (2002). *Experimental and quasi-experimental designs for generalised causal inference*. Boston: Houghton Mifflin.
Simon, H. A. (2012). *Models of discovery: And other topics in the methods of science* (Vol. 54). Berlin: Springer Science & Business Media.
Sokolowski, A. (2015). The effects of mathematical modelling on students' achievement, a meta-analysis of research. *IAFOR Journal of Education, 3*(1), 93–114.
Sokolowski, A. (2018a). Modeling acceleration of a system of two objects using the concept of limits. *Physics Teacher, 6*(1), 40–41.
Sokolowski, A. (2018b). *Scientific inquiry in mathematics—Theory and practice: A STEM perspective* (1st ed.). New York: Springer.
Tipler, P. A, & Mosca G. (2007). *Physics for scientists and engineers, Vol. 1, 6th: Mechanics, oscillations and waves, thermodynamics*. New York: Freeman Company.
Tytler, R., Prain, V., & Peterson, S. (2007). Representational issues in students learning about evaporation. *Research in Science Education, 37*(3), 313–331.
Uhden, O., Karam, R., Pietrocola, M., & Pospiech, G. (2012). Modelling mathematical reasoning in physics education. *Science & Education, 21*(4), 485–506.

Chapter 8
Quantitative Reasoning and Its Rôle in Interdisciplinarity

Robert Mayes

Abstract The Real Science, Technology, Engineering Mathematics (STEM) Project was conducted in middle schools and high schools in Georgia, USA. The project supported the development of interdisciplinary STEM modules and courses in over 20 schools. A project focus was development of five 21st century STEM reasoning abilities. In this chapter, I provide classroom activities from the Real STEM project that exemplify each form of reasoning: complex systems; model-based; computational; engineering design-based; and quantitative reasoning. Quantitative reasoning plays a critical rôle in authentic real-world interdisciplinary STEM problems, providing the tools to construct data informed arguments specific to the problem context, which can be debated, verified or refuted, modelled mathematically and tested against reality. Yet quantitative reasoning is often misrepresented, underdeveloped, and ignored in STEM classrooms. The chapter finishes with a discussion of the impact of Real STEM.

Keywords Quantitative reasoning · STEM reasoning · Authentic teaching · Learning progression

8.1 Introduction

Interdisciplinary Science, Technology, Engineering, and Mathematics (STEM) teaching and learning is a national obsession in the United States. There are calls to have STEM integrated into all schools from elementary level through to university. Why? First, there is the economic driver of increasing the number of students pursuing STEM areas to address growing STEM workforce needs. Second, there is the desire for STEM literate citizens, who can make informed decisions about grand challenges facing the next generation, challenges such as global climate change, clean water and the future of energy. Third, there is the proposed positive benefit of

R. Mayes (✉)
Georgia Southern University, Middle Grades and Secondary Education PO Box 8134, Statesboro, GA 30460-8134, USA
e-mail: rmayes@georgiasouthern.edu

B. Doig et al. (eds.), *Interdisciplinary Mathematics Education*,
ICME-13 Monographs, https://doi.org/10.1007/978-3-030-11066-6_8

increasing student engagement, and persistence, in STEM areas through authentic teaching approaches, that provide hands-on, collaborative opportunities for students. The National Science Teachers Association, and the National Council of Teachers of Mathematics, promote standards that support STEM, including the Next Generation Science Standards (NGSS Lead States, 2013) and the Common Core State Standards for Mathematics (National Governors Association, 2010).

Mathematics is fundamental to interdisciplinary STEM, providing the processes to quantify a science problem, analyse engineering designs, and model large data sets. Applying mathematics to real-world interdisciplinary STEM problems requires more than knowledge of isolated mathematical algorithms. It requires the student to quantify the STEM context, and to select the appropriate mathematical tool for a given problem. The ability to apply mathematics within a real-world context is at the core of quantitative reasoning. Unfortunately, many students observed in our Real STEM Project, did not have good quantitative reasoning skills, not even those who were skilled at manipulation and calculation.

From Spring 2013 through to Spring 2017 the Real STEM Project supported over 39 teachers in 20 partner schools in creating and offering interdisciplinary STEM research, and design, experiences for students from age 12 to 18. The project advocated that quantitative reasoning is essential in integrating interdisciplinary STEM into school curricula. The Real STEM project went further by identifying five 21st century STEM reasoning modalities, that are of high demand in the work force, and support being a STEM literate citizen. These five STEM reasoning modalities are: complex systems reasoning, scientific model-based reasoning, technologic computational reasoning, engineering design-based reasoning, and mathematical quantitative reasoning. In this chapter, I begin by discussing what interdisciplinary STEM teaching and learning means, present these five STEM reasoning modalities and provide authentic problem-solving situations in which these reasoning modalities were explored by students in partner schools. This chapter structure enables me to illustrate problem-based learning and place-based education in authentic settings as experienced by Real STEM project teachers and students. Teacher (n = 39) and student responses (n = 898) to increased engagement through interdisciplinary STEM problems was very positive and the chapter concludes with a discussion of how to take such work forwards.

8.2 Interdisciplinary STEM: Authentic Teaching and Reasoning Modalities

STEM is the collective study of science, technology, engineering, and mathematics, with the goal of equipping students with the knowledge and skills to solve tough problems, gather and evaluate evidence, and make sense of information (U.S. Department of Education, 2015). STEM is, first and foremost, interdisciplinary. The term STEM is not needed if you have a great science programme, just call it a great science programme. STEM occurs when two, or more, of the areas of science, technology,

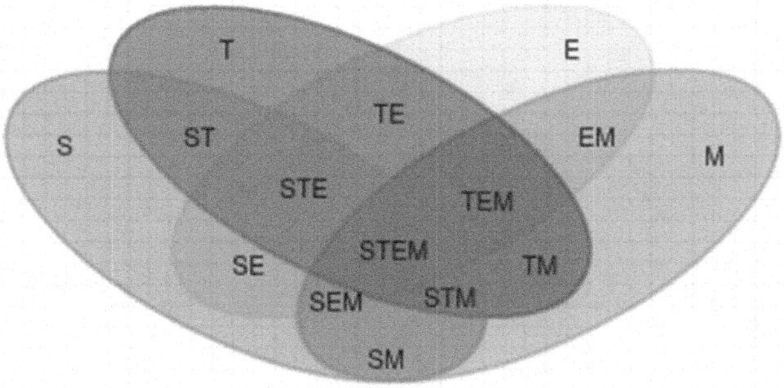

Fig. 8.1 STEM occurs in the intersections of the Venn diagram, not the single set spaces

engineering, and mathematics are brought to bear on a problem (Fig. 8.1). STEM engages students in authentic learning, by engaging them in real-world problems that are student centred and, when possible, tied to the student's context.

Real STEM project was based in what is known as Problem-Based Learning (PBL), a learner-centred approach that empowers learners to conduct research, integrate theory and practice, and apply knowledge and skills to develop a viable solution to a defined problem (Savery, 2006). Long-term retention, skill development, and student and teacher satisfaction, have been found to be benefits of problem-based learning when compared with traditional forms of instruction (Strobel & van Barneveld, 2009). Statistically significant gains in achievement, have been observed for middle school science students, experiencing science in a problem-based learning format (Williams, Pedersen, & Liu, 1998). Place-Based Education (PBE) uses the environment as a context for learning, and allows student input on the selection of the problem to be researched. Studies have found that PBE resulted in students who scored higher on standardised tests in reading, writing, mathematics, science, and social studies (Lieberman & Hoody, 1998; Bartosh, 2003; NEETF, 2000). Other results, from these PBE studies, indicated that students improve overall Grade Point Average (GPA), stay in school longer, and receive higher than average scholarship awards. Authentic learning has a number of qualities, including use of: real-world relevance, ill-defined problems, sustained investigation, collaboration, interdisciplinary perspective, and time for reflection.

The more student centred and problem-driven the STEM task is, the harder it is to target specific science, or mathematics, concepts. So, why should science and mathematics teachers implement authentic STEM tasks in their classrooms? Potential affective outcomes of authentic learning are increased student engagement and persistence, but what are the learning outcomes? Important learning outcomes include critical thinking, problem solving, and the ability to reason. Literally the student should gain a better understanding of how a scientist, computer scientist, engineer, and mathematician, solve a problem. Five STEM reasoning modalities are of high demand in the work force: complex systems reasoning, scientific model-based reasoning, technologic computational reasoning, engineering design-based reasoning,

Fig. 8.2 Starlings flocking in Rome https://youtu.be/V-mCuFYfJdI

and mathematical quantitative reasoning. An exemplar for each reasoning modality is provided below, together with a focus on the rôle that quantitative reasoning plays in each.

8.2.1 Complex Systems Reasoning

Birds flocking is an exemplar of a complex system. Starlings flocking in Rome (Fig. 8.2) organise themselves based on simple interactions between birds (the agents), the result of which, are beautiful emerging patterns from random bird interactions. Flocking is an adaptation that, among other things, confuses predators such as the Peregrine Falcon.

Complex systems reasoning is the ability to analyse problems, like flocking behaviour, by recognising complexity, patterns, and interrelationships within a system featuring a large number of interacting components (agents, processes, etc.) whose aggregate activity is non-linear (not determined from the summations of the activity of individual components) and typically exhibits hierarchical self-organisation under selective pressures (Holland, 1992). Complex systems are characterised by a number of elements including: agent-based reasoning, where individual system elements and their interaction are considered; complexity, with a multi-scale hierarchical organisation of smaller systems within larger ones; emergence of patterns, from random interaction of agents; and self-organisation, to adapt to the environment. If the teacher's goal is real-world problem-driven experiences for students, then it is likely the problem lies within a complex system, such as a biological ecosystem.

Simulations of flocking allow students to explore and discover the three simple rules of flocking: Cohesion (steering to move toward the average position of local flockmates); Alignment (steering toward the average heading of local flockmates); and Separation (steering to avoid crowding local flockmates). A flocking simulation is provided in NetLogo, an open software programmable modelling environment, for simulating natural and social phenomena (Wilensky & Resnick, 1999). You can download NetLogo for free and it comes with an extensive library of simulations and lessons: https://ccl.northwestern.edu/netlogo/download.shtml.

The flocking simulation can be found in NetLogo by selecting File—Models Library—Biology—Flocking—Open. The simulation allows the student to vary population, vision, and the three rules of flocking. Running the simulation shows students the emerging behaviour of flocking, without the benefit of a lead bird. The concept of a system organising, without a leader, is surprising to students, and represents a central concept of complex system reasoning: self-organisation due to agent interaction.

So, where is the interdisciplinary STEM in the flocking example? Where is the mathematics? Science is evident, with biology, and environmental science, serving as the driver for the problem, while physics can be invoked through a study of flight. Science is one of the most common drivers in school interdisciplinary STEM tasks. If students are left to their own devices in exploring the simulation, the result is often a qualitative science account of flocking. But there is much more to the problem, if students are directed to view the problem through multiple STEM lenses, and to provide data-based arguments supporting their analysis. Technology is present in use of the simulation, the programming behind the simulation, and in engaging students in extending the model through programming. Engineering design, can be incorporated by having students engineer flying machines and compare them to the birds natural design. Mathematics underlies the development of the simulation. Quantitative reasoning, about the rules governing the interaction of the birds, provides for rich mathematical discussion.

Questions to ask the students to explore include:

- What is the most efficient minimum separation for flocking? This evokes distance, measurement, and inequality (distance to nearest bird < minimum separation).
- What is the effect of changing the maximum separation turn degrees? This evokes geometry.
- How do I calculate alignment of birds? This evokes trigonometry, specifically use of the arctangent.
- How do I calculate coherence of birds? This evokes use of sine and cosine.

8.2.2 Model-Based Reasoning

This occurs when students construct scientific models in order to explain observed phenomena (MUSE, 2015). Common problems in teaching the scientific method are that it becomes an algorithm that is followed linearly, and the result of the research is not related back to the real-world context. For example, a study of a local pond includes collection of dissolved oxygen data. The students collect the data, create a plot, and perhaps even determine trends, but they fail to discuss what the level of dissolved oxygen means for the pond ecosystem. The question of what type of fish could live in the pond, if dissolved oxygen is varying between 2 and 6 mg/litre is never addressed. Model-based reasoning is characterised by: development of a scientific model; revision of the model; determination of the acceptability of the model, in explaining all observations; predicting behaviour of the system; being consistent with other science; empirical assessment (the model explains the data and predicts future results); and conceptual assessment (the model fits with other accepted models).

Model-based reasoning can be elicited by having students engage in the seven step, model-based, reasoning process (Schwarz et al., 2009).

(1) Students observe an anchoring phenomenon of which they do not have a complete understanding.
(2) Students construct a model that expresses an idea or hypothesis about what is happening. This conceptual model includes a picture of the phenomenon with components identified, connections between components represented, and variables quantified when possible, which engages the student in quantitative reasoning.
(3) The students empirically test the model by determining if it reflects the reality or the phenomenon they observed.
(4) They evaluate the model against any data they have collected.
(5) The students test their model against other scientific ideas, laws, or theories, to see if it is consistent with known science.
(6) Once the original model is complete, the students are asked to revise the model to fit new evidence and known laws of science.
(7) Finally, they apply the model to make predictions and explain the phenomenon.

The phenomenon of electrolysis of water, provides a good exemplar for model-based reasoning. Set up the experiment as in Fig. 8.3, without discussing the concept prior to student observation of the phenomenon. Have students go through the seven step model-based reasoning process (Schwarz et al., 2009) described above. Pure water is an insulator, and the electrolysis may proceed too slowly, so add salt to speed up the reaction. Using NaCl as an electrolyte results in some impurity in the form of chlorine gas at the anode, but that should not be important for the purpose of this demonstration. Bubbles will form at the tips of pencils immediately. Oxygen gas bubbles (O_2) will form at + electrode (anode). Hydrogen gas bubbles H_2 will form at − electrode (cathode). The amount of H_2 will be twice the amount of O_2.

In this experiment science is again the driver, with chemistry and physics being evident. Technology, as computational reasoning, does not play a rôle. Engineering design, could be integrated into the experiment, by asking students to design an apparatus that allows them to collect and measure the gases being generated. Expect that student's initial models will be qualitative accounts, which include the

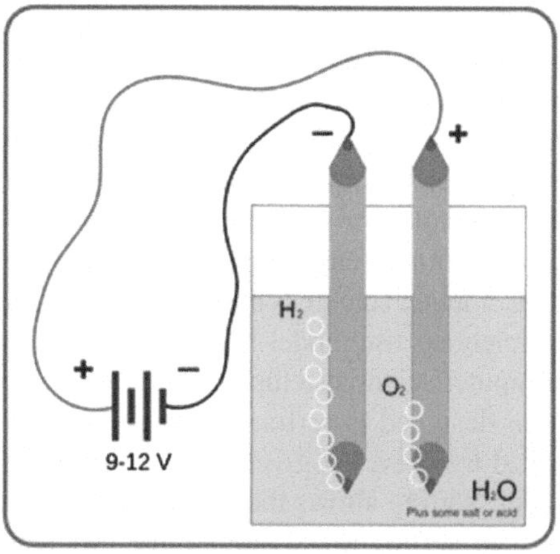

Fig. 8.3 Electrolysis of water experiment

key components of the experiment, provide an explanation of what is generating the bubbles, and may label the gases being generated. Most likely, it will not include a quantitative account, unless specifically prompted. Quantitative analysis of the electrolysis of water phenomenon includes rate of gas production, amount of gas produced, and balancing chemical equations that represent the reaction. The authentic, real-world, aspect of the experiment is the production of hydrogen as a renewable energy resource. This elicits questions of the feasibility of large scale production, and the cost of transforming electric energy into hydrogen, as a fuel source. Both of these provide further opportunities for integrating quantitative reasoning into the task.

8.2.3 *Computational Reasoning*

Computational reasoning is an analytical approach grounded in the computer sciences, that includes a range of concepts, applications, tools, and skills, that allow us to solve problems strategically, design systems, and understand human behaviour, by following a precise process, that engages computers to assist in automating a wide range of intellectual processes (Wing, 2006). Computational reasoning and large database analysis, are considered new paradigms of science, expanding beyond the traditional experimental and theoretical science paradigms. Computational reasoning is often viewed by teachers as programming, but, there is more to how computer scientists reason, than programming. Elements of computational reasoning include: abstraction by stripping a problem down to its bare essentials, then transferring the problem solving process to a wide variety of problems; algorithm design; automation of repetitive tasks to perform them quickly and efficiently; decomposition of a problem into steps that are implementable by machines; parallel processing; programming, simulation, and modelling; visualisation of large data sets; and capture and curation of data.

Computer programming is the most common activity implemented in classrooms to engage students in computational reasoning. Object oriented programming languages, such as Scratch (https://scratch.mit.edu/), make programming more approachable for students even at the 9–11 years of age level. However, teaching computational reasoning does not require expensive technology, or programming.

Computer scientists have their own ways of solving real-world problems. As much as scientists use model-based reasoning, and engineers use design-based reasoning, computer scientists employ algorithmic reasoning. An algorithm is a set of instructions designed to perform a specific task. For example, computer scientists are often asked to sort a list given some criteria, for example putting names into alphabetic order. Sorting allows the user to find a name more efficiently. There are many methods for sorting a list, some more efficient than others. The following sorting task engages students in creating their own sorting algorithm, and testing it against others, to see which sorts most efficiently. Provide the students with 8 film canisters of the same size but with different weights (e.g. filled with different amounts of sand) and a simple balance scale. Determine the best method of sorting the 8 containers provided so they are in order from lightest to heaviest. Make a comparison of efficiency with another group by swapping sorting algorithms and seeing who can sort the containers in the fewest moves. Clarity of the algorithm will be essential for others to complete the sort, so students need to determine if the algorithm was detailed, and clear enough, to follow easily.

This is a great computational reasoning task, but is it interdisciplinary STEM? The task does not include science, but engineering design could be incorporated if students are asked to engineer a measuring device for sorting the containers. Technology is emphasised by requiring algorithmic reasoning, and exploring different sorting algorithms. There are a number of sorting algorithms students might discover, such as the selection sort (lightest weight is found and removed, then repeat with remaining canisters) or the insertion sort (with each sort the new canister is placed in the appropriate position in the previously sorted canisters). Other sorts include the bubble sort, quick sort, and the mergesort. But where is the mathematics in the sorting task? The quantitative reasoning arises in the efficiency count. Ask students to provide a quantitative account of the efficiency of their algorithm. This engages the students in the discrete mathematics area of combinatorics (counting without counting). For example, for the insertion sort the students can begin by calculating the efficiency for an increasing number of containers. For 3 containers it is a maximum of 2 measures, for 4 containers 5 measures, for 5 containers 9 measures. A pattern emerges, where for n containers ($n > 2$) the number of measures is the sum of the integers from 2 to $n - 1$. Adjusting the formula for the sum of the first n integers, to account for starting at 2 (subtract 1) and ending at $n - 1$, the number of measures for n canisters would be given by the formula:

$$\frac{(n-1)n}{2} \tag{1}$$

In addition, a comparison of efficiency for different sorting algorithms provides the opportunity for integration of algebraic inequalities.

8.2.4 Engineering Design-Based Reasoning

Engineering design-based reasoning is the ability to engage in the engineering design process, thereby using a series of process steps to come up with a solution to a problem (PLTW, 2017). Many times the solution involves designing a product that meets certain criteria and, or, accomplishes a certain task. There are a variety of engineering design process models, but most have variations of the following steps:

(1) Define the problem including determining criteria the design must meet and the constraints on the design;
(2) Research the problem;
(3) Brainstorm solutions;
(4) Choose the best design;
(5) Build a prototype;
(6) Test the prototype; and
(7) Redesign.

As with the scientific method, it is important to stress that the design process is not a linear process, but a circular one which often requires jumping back to previous steps in the process. Engineering design tasks and scientific experiments were the two most common problem drivers in our observations of Real STEM project classes.

Alternative energy design problems provide a good engineering context. For example, designing an efficient wind turbine blade, provides an excellent interdisciplinary STEM task. A base for the wind turbine can be provided, including a stand, motor, and wiring (Fig. 8.4). The U.S. Department of Energy has detailed plans for a wind turbine base in Building the Basic PVC Wind Turbine (http://www1.eere.energy.gov/education/pdfs/wind_basicpvcwindturbine.pdf).

A student is only responsible for the design of the wind turbine blade. A variety of materials, for blade construction, can be provided for students to choose from, or the teacher can allow the students to forage for materials. Students work in teams to design blades from selected materials, and mount them on a hub that can be connected to the wind turbine base. The engineering design process guides all aspects of the blade development, including the critical component of identifying criteria, and constraints, for blade construction. Constraints of materials used to create blades provide an opportunity to integrate materials science into the task. Criteria can be set by students, and include outcomes, such as continuous blade rotation for at least a minute, blade rotation speed, or a minimum energy production requirement. A multimeter can be used to measure volts produced. Students keep a design notebook, where they log each step in the engineering design process.

Redesign is an essential part of engineering design. Students can consider a number of variables that may have an impact on blade performance: blade length and

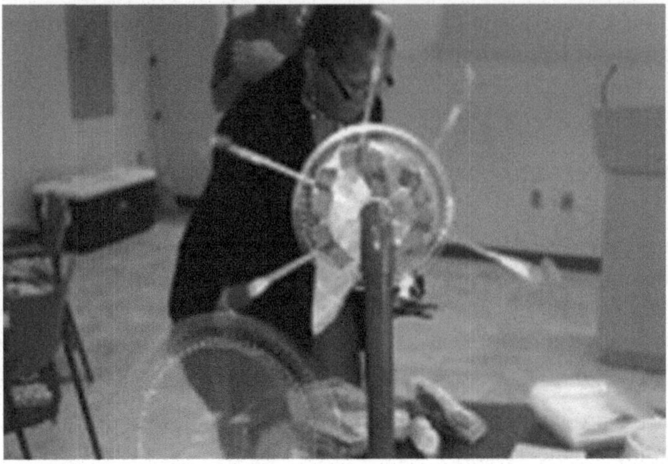

Fig. 8.4 Wind turbine base and set up for testing blades

$$P = \tfrac{1}{2}\,\rho\,\pi\,r^2 V^3$$

P = Power in the Wind (watts)

ρ = Density of the Air (kg/m^3)

r = Radius of your swept area (m)

V = Wind Velocity (m/s)

π = 3.14

Fig. 8.5 Power in the wind formula

pitch, number of blades, material used in blades, including smoothness of surface, and blade shape. Engineering design tasks have great potential for interdisciplinary STEM learning, but observations of Real STEM classes indicated that the potential was often not realized. If students are not held to rigorous engineering design standards, then the tasks may devolve into a trial and error mode, where there is no observable science, technology, or mathematics. The wind turbine blade task can engage students in the earth systems science topic of wind and weather, as well as physics of power. The task supports the use of technology including circuits, motors, and a multimeter. It also has plentiful mathematical applications. A simple equation for power can be used to provide a quantitative reasoning aspect to the problem (Fig. 8.5). With this equation students can determine the power generated by a typical house fan with wind velocity $V = 5$ m/s (metres per second), density of air $\rho = 1.0$ kg/m^3 (kilograms per cubic metre), and radius of swept area $r = 0.2$ m.

8.2.5 Quantitative Reasoning

Having discussed the integration of mathematics into the other four STEM reasoning modalities, now I focus on quantitative reasoning itself. The Culturally Relevant Ecology, Learning Progressions, and Environmental Literacy project (Mayes, Peterson, & Bonilla, 2012, 2013) developed a definition of quantitative reasoning, as well as a learning progression proposing a trajectory of QR development across ages 12–18.

Quantitative Reasoning (QR), in context, is mathematics and statistics, applied in real-life, authentic, situations, that have an impact on individual's life as a constructive, concerned, and reflective citizen. QR problems are context-dependent, interdisciplinary, open-ended, tasks that require critical thinking and the capacity to communicate a course of action (Mayes et al., 2013).

Once QR was defined, the research team began to construct, based on the literature and professional experience, a framework for QR that would be evaluated through development of a learning progression. The four key components of QR in the framework and key researchers' work upon which they were determined are:

1. Quantification Act (QA): The mathematical process of conceptualizing an object, and an attribute of it, so that the attribute has a unit measure (Thompson, 2011; Dingman & Madison, 2010).
2. Quantitative Literacy (QL): The use of fundamental mathematical concepts in sophisticated ways for the purpose of describing, comparing, manipulating, and drawing conclusions, from variables developed in the quantification act (Steen, 2001; Madison, 2003; Briggs, 2004).
3. Quantitative Interpretation (QI): Ability to use models to discover trends, and make predictions (Madison & Steen, 2003; Thompson & Saldanha, 2000).
4. Quantitative Modelling (QM): The ability to create representations that explain a phenomenon, and to revise them based on fit to reality (Duschl, Schweingruber, & Shouse, 2007; Schwarz et al., 2009; Lehrer, Schauble, Carpenter, & Penner, 2000).

A learning progression is a set of empirically grounded, and testable, hypotheses about how, with appropriate instruction, students' understanding of, and ability to use, core scientific concepts, explanations, and related scientific practices, grow and become more sophisticated over time (Corcoran, Mosher, & Rogat, 2009). Learning progressions provide levels of understanding through which students develop mastery of a concept over an extended period of time, such as over six years from ages 12–18. The QR learning progression is conceptualised as having four levels: the lower anchor, upper anchor and two intermediate levels of understanding. The lower anchor is grounded in data collected on 12 year olds understanding of QR (Mayes et al., 2014b). The upper anchor is based on expert views of what a scientifically literate citizen, who is well versed in QR, should know, and be able to apply by age 18. A learning progression defines progress variables, which are essential categories for the overall concept across which the levels are established. The progress variables for the QR learning progression were drawn from the four components in the QR framework (Mayes, Forrester, Christus, Peterson, & Walker, 2014a). The Quantification Act, and Quantitative Literacy, components were combined under Quantification Act (QA), with the expectation that once a variable is conceptualized, then fundamental mathematical concepts allow one to compare, contrast, manipulate, and combine the variables to form mathematical expressions. Quantitative literacy is essential to moving, from quantification, to building and interpreting models. This reduction of the framework left three progress variables in the QR learning progression: Quantification Act (QA), Quantitative Interpretation (QI), and Quantitative Modelling (QM).

Finally, each of the progress variables were elucidated by identifying a collection of elements, that indicate essential capabilities within the categories that were determined through student interviews, and tested, throughout development of the learning progression:

- Quantification Act Elements: Variation, Quantitative Literacy, Context, Variable
- Quantitative Interpretation Elements: Trends, Predictions, Translation, Revision
- Quantitative Modelling Elements: Create model, Refine model, Reason with model, Statistical analysis.

For a detailed presentation of the learning progression see Mayes et al. (2014a).

8.2.5.1 Quantitative Reasoning Examples

Mathematics is not typically the driver for STEM in schools. In STEM, S, and E, are the most common drivers, with T, and M, playing support rôles. Can mathematics be the driver in STEM? Certainly, verification of a mathematical statement, or a mathematical argument supporting a conjecture, can be a driver in STEM, if the conjecture is connected to a real-world STEM application. The difficulty is engaging a broad range of students in such mathematical discourse. Quantitative reasoning provides the opportunity for mathematics to play a more central rôle in STEM.

For authentic real-world interdisciplinary STEM problems, the quantification act is the ability to mathematise the problem, moving from a qualitative account to a quantitative description, by establishing quantitative variables, connecting the variables through exploring covariation, and building algebraic expressions. Quantitative modelling is the creation and refinement of a model, reasoning with mathematical models, and the use of statistical inference, to test hypotheses springing from analysis of data gathered on the problem. Quantitative interpretation is using a model to determine trends and make predictions, revision of models to fit reality, and the translation between multiple models of the same problem. Here I use the topic of sustainable energy, and environmental impacts, as a context for exploring the rôle of quantitative reasoning in STEM (Mayes & Myers, 2014). The amount of data, and variety of representations, in the area of energy challenge quantitative interpretation abilities.

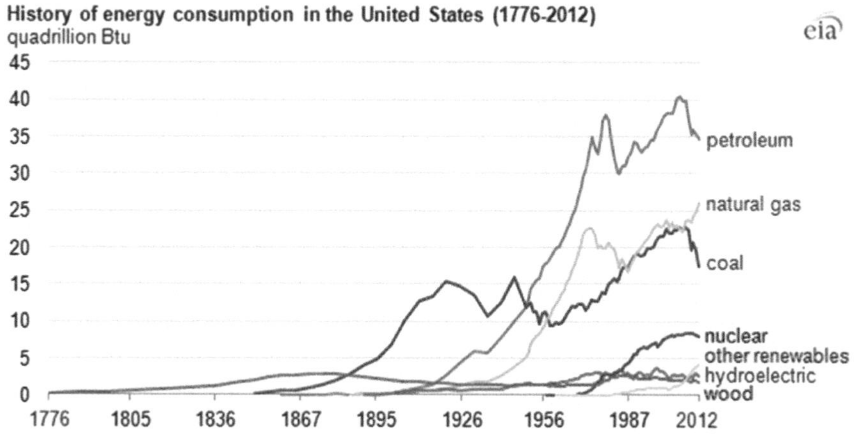

Fig. 8.6 Energy models (Mayes & Myers, 2014). U.S. Energy History. *Source* U.S. Energy Information Administration (2012)

Quantitative reasoning is recognised as an essential component for making informed decisions (Mayes et al., 2012). Quantitative reasoning, moves past quantities and values, to conceptualise and interpret, the relationships and contexts defining them (Thompson, 1993; Ramful & Ho, 2015). Within an interdisciplinary framework, quantitative reasoning, is envisioned as the application of mathematical concepts and models across domains to discover trends, and make inferences and predictions (Mayes & Koballa, 2012; Elrod, 2014). Further, quantitative reasoning distances itself from traditional mathematics, through its emphasis on ill-defined, open ended, real world problems (Mayes et al., 2012; Elrod, 2014). Unlike traditional mathematics, which places emphasis on calculations, and manipulations of abstract representations, quantitative reasoning is distinct in its emphasis on the underlying meaning of mathematical functions, and its application to authentic real-world problems (Elrod, 2014). Quantitative reasoning problems are context dependent, interdisciplinary, and open-ended, tasks that require critical thinking and the capacity to communicate a course of action. The energy exemplar, below, elucidates major rôles for quantitative reasoning in interdisciplinary STEM. Consider the representations in Fig. 8.6. Students need to understand the variable attributes and measures (QA) before they can interpret a model (QI).

Modelling data, and testing statistical hypothesis (QM), are critical for many real-world STEM problems. Given the data on U.S. Oil Consumption and Production in Table 8.1, there are a number of quantitative analyses that can be performed in analysing these data.

Table 8.1 U.S. Oil Consumption and Production, 1990–2018 (Year 1 is 1990)

Year	Production	Consumption	Year	Production	Consumption
1	7.36	16.99	16	5.18	20.80
2	7.42	16.71	17	5.09	20.69
3	7.17	17.03	18	5.08	20.68
4	6.85	17.24	19	5.00	19.50
5	6.66	17.72	20	5.35	18.77
6	6.56	17.72	21	5.48	19.18
7	6.46	18.31	22	5.65	18.88
8	6.45	18.62	23	6.49	18.49
9	6.25	18.92	24	7.47	18.96
10	5.88	19.52	25	8.76	19.11
11	5.82	19.70	26	9.41	19.53
12	5.80	19.65	27	8.85	19.63
13	5.74	19.76	28	9.35	19.97
14	5.65	20.03	29	9.91	20.30
15	5.44	20.73			

Production and consumption unit is million barrels per day ($\times 10^6$). U.S. Energy Information Administration https://www.eia.gov/outlooks/steo/data/browser

8.2.5.2 Descriptive Statistics Analysis

Use Excel to find measures of centre (mean, median, mode) and spread (range, standard deviation). What do these descriptive statistics tell you about the data sets?

Should you use the mean, median, or mode for this data set? Should you use the range, or standard deviation? Why? Construct a histogram of the data, to explore issues of data distribution type, related to which measure of centre and spread to use. Statistical display:

- QM Model—Data Display: Which is the best data display to use for these data (frequency table, bar chart, histogram, pie chart, scatter plot, dot plot, stem and leaf plot, box and whisker plot)? Why?
- QM Trends and Predictions: Use the data display you selected to discuss trends in the production and consumption data. Make a prediction of production and consumption in 2020.

Modelling:

- QM Mathematical Model: Create a mathematical model for production by year, using a line, or curve, of best fit. Use the model to extend the discussion of trend, and verify your prediction.
- Now find the line of best fit for the consumption by year. Predict consumption in 2020. Extension: Attempt a curve of best fit, for example a parabola. Does it fit

Table 8.2 Selecting a statistical test

Test	Hypothesis	Indep't Var'ble	Dep't Var'ble	Co-variates	Contin. or Cat. Ind'p't Var	Contin' or Cat. Dep't Var	Dist'n
t test	Grp Comp	1	1	0	Cat.	Cont.	Normal
ANOVA	Grp Comp	1or more	1	0	Cat.	Cont.	Normal
Chi-sq.	Grp Comp	1	1	0	Cat.	Cat.	Non-normal
Pearson R	Relate Var's	1	1	0	Cont.	Cont.	Normal

better than the line of best fit? Is the additional complication of a quadratic model worth it?

Hypothesis testing:

- QM Hypothesis Testing: Is the difference between production and consumption significant? We can test that question by comparing the means of the production and consumption data sets. First, examine a visual display of the two data sets, to see if they appear to be significantly different. Construct box and whisker plots for both production and consumption.

  An easy online boxplot tool is the Boxplot Grapher (http://www.imathas.com/stattools/boxplot.html).

  Do the plots support the hypothesis that there is a significant difference? Can we say the difference is significant using only a visual display?
- QM Statistical Hypothesis Testing: While comparing box and whiskers plots provides some intuition about differences in data sets, determining if there is a statistically significant difference requires conducting a formal statistical analysis. First, determine the best statistical test to use to assess the null hypothesis that the two data sets are not statistically different. Table 8.2 identifies four basic statistical tests, and the conditions under which they should be used. Which works best for this problem?
- The descriptive statistics, called for above, provide some information on which to conjecture about the type of distribution criteria. Use histograms for the two data sets for a visual representation of distribution type (normal or nonnormal). Which statistical test is best to use considering what you know about the production and consumption data sets?
- Use Excel to run the best statistical test for the null hypothesis. Use the Data—Data Analysis—t-Test: Two-Sample assuming Unequal Variances (NOTE t-test has several versions, this is the best for this data). Run with an alpha level of 0.05 (5% risk that we accept a difference that does not exist, that is the null hypothesis true

Fig. 8.7 Type I and Type II errors

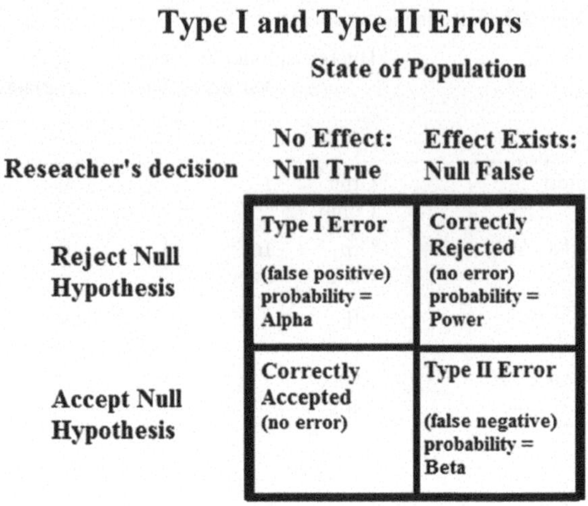

but we say it is false). Interpret the resulting table. Is there a statistically significant difference between the data sets?

8.2.5.3 Inferential Analysis—Hypothesis Testing

It is important that students understand the concept behind hypothesis testing. We begin with a null hypothesis, that there is no difference between the data sets. We set an alpha level, typically $\alpha = 0.05$, which indicates the maximum risk we are willing to take that any observed differences are due to chance. So for $\alpha = 0.05$ we are willing to risk 5% of the time that we say there is a significant difference when there is not (Fig. 8.7). This is called a Type I error, where we have a false positive which is the worst possible outcome. We can also be in error when there is an effect but the test does not pick it up. This is called a Type 2 error, but we have a false negative which is at least more conservative, and therefore, less of a concern.

- State the Type 1 and Type 2 errors in terms of the production-consumption comparison.

Unless we have a good reason to believe that prior to the experiment the relationship will occur in one direction, such as that consumption will always exceed production, then we use a two-tailed test. If we do have a sense of direction for the outcome we use a one-tailed test. Let's use a two-tailed test for our comparison. Hypothesis testing is a probability game that indicates if we should accept or reject the null hypothesis (Fig. 8.8). If the probability value p is high then the null hypothesis is likely true and we do not reject it. If the probability value is less than $\alpha =$ 0.05 then it is highly unlikely any difference is due only to chance (fewer than 5% of studies would result in the difference due to random sampling error only) so we reject the null hypothesis.

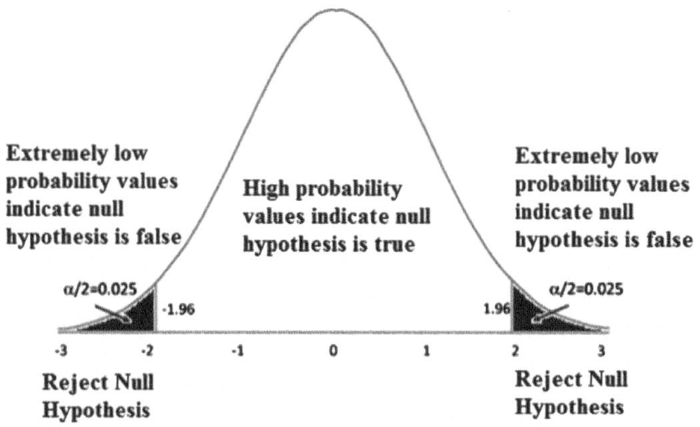

Fig. 8.8 A two-tailed test of a null hypothesis at an alpha level of 0.05

- State the above, with reference to the production-consumption comparison.

This concludes the presentation of examples of quantitative reasoning. The Real STEM Project classroom observations revealed that students struggled with QA and QI, and that teachers rarely engaged them in QM. Too often, interdisciplinary STEM tasks assigned by teachers, did not move the students beyond qualitative science descriptions or engineering designs lacking in quantitative analysis.

8.2.6 Evaluation

Teacher (n = 39) and student responses (n = 898) to increased engagement through interdisciplinary STEM problems was very positive. Teacher focus groups indicated positive interest and activity in developing STEM partnerships with businesses and research institutes. The teachers had areas of concern with implementing the STEM reasoning modalities, and sustaining collaboration with teachers in other STEM areas. The middle schools (ages 12–14) found it easier to have teachers collaborating on STEM research and design courses, due to the cross disciplinary team structures that exist in many middle schools, and the availability of flexible courses as a natural place to implement STEM courses. For example, Connections Courses, which provide opportunities for middle grade students to explore high school career pathways, were the common vehicle for establishing an interdisciplinary STEM course at the middle school level. The subject area silo structure of high schools, and the barrier of developing and staffing new courses in STEM, made it more of a challenge for high schools (ages 15–18) to sustain interdisciplinary STEM courses. In order to overcome the teacher collaboration and structure issues, of implementing interdisciplinary STEM programmes, it is essential to have administrative support and participation. The most successful high school STEM courses thrived when given administrative support.

Student surveys were conducted to determine student reaction to the STEM courses. Overall, the results indicated that students expressed statistically significant increases in Intrinsic Motivation, Self-Management and Self-Regulation, and Intent to Persist in STEM. The largest student gains observed were in the Intrinsic Motivation construct. For example, before taking the STEM courses 54% of students said that they enjoyed challenging classwork; after completing the courses 75% of students agreed that they preferred classwork that was challenging. Likewise, before taking the courses 62% of students agreed that the content they were learning could be used in other classes. After the courses, 81% of the students felt that they would be able to use what they learned in their other classes. A second survey examined student (1) interest in STEM fields, (2) confidence in their ability to perform academically in STEM fields, (3) feelings about the importance of understanding STEM, (4) interest in taking classes and pursuing post-secondary education in STEM fields, and (5) interest in STEM careers. There were significant differences in all five categories, supporting improved student attitudes and beliefs, upon completion of a STEM course.

8.3 Conclusion

The push to incorporate interdisciplinary STEM into existing science and mathematics classes, as well as for development of new STEM research and design courses, provides an excellent opportunity for interdisciplinarity for mathematics. STEM problems are real-world, complex, and require cross-disciplinary applications. Quantitative reasoning is a natural fit for such problems, consisting of the tools and concepts supporting quantification, interpretation, and modelling of STEM problems. The challenge for STEM in general, and mathematics specifically, is that quantitative reasoning abilities are not well developed in most students. We need to develop mathematical reasoning across STEM in an interdisciplinary manner.

The Real STEM project provides a model for developing and integrating interdisciplinary STEM courses into traditional middle schools and high schools. In the USA, intensive interdisciplinary STEM programmes are often the province of specialized STEM magnet schools or academies. The demand for STEM understanding far exceeds these specialized schools, both for workforce needs, and for STEM literate citizens. But, there are extensive barriers to integrating STEM into traditional middle schools and high schools, including curricula guided by excessive high stakes testing, extensive curriculum implementation guidelines, that limit flexibility in both topics taught and order in which they are presented, teachers' fear of going beyond their disciplinary boundaries, inflexible school schedules, that inhibit cross-disciplinary planning time, and lack of administrative support for interdisciplinary STEM and authentic teaching. So what can stakeholders take from the Real STEM project?

Practitioner stakeholders can use the Real STEM tenets as a guide to implementing interdisciplinary STEM in their classrooms. ICARE is an acronym for the key tenets:

- Interdisciplinary STEM that integrates the four STEM subjects across mathematics, science, engineering, and technology courses.
- Collaboration both within schools, through interdisciplinary STEM Professional Learning Communities (PLC), and external to the school, though partnerships that bring STEM experts into the classroom from the community, business, and industry, and research institutes such as universities and government research entities.
- Authentic teaching strategies, that engage students in real-world problems, and provide opportunities for student-centred research and design performance tasks.
- Reasoning in STEM that moves beyond entertaining activities, to performance tasks that reflect understanding and reasoning.
- Education for Understanding, that identifies enduring understandings, and essential questions, that motivate students to engage in developing and demonstrating deeper conceptual understanding.

Practitioners should focus on providing opportunities for model-based reasoning, design-based reasoning, and quantitative reasoning. All of these permeate the four subject areas of STEM, and when applied in real-world contexts, provides the opportunity to incorporate social sciences. Implementing the ICARE tenets requires teachers and administrators to work together to overcome barriers, such as, common PLC planning time, development of community partnerships, and flexibility in shuffling curriculum to allow for collaborative lessons across subject areas.

Policy-maker stakeholders should take notice of our research, that indicated the Real STEM project's impact on improving positive student engagement and the students' struggle with interdisciplinary STEM reasoning. Reduce excessive testing and curricular control, allowing teachers the flexibility to use authentic teaching strategies to improve student understanding of STEM.

For researchers, the project outcomes include a Quantitative Reasoning Learning Progression, a diagnostic assessment of interdisciplinary STEM for middle and high school grades (ages 12–18), student and teacher attitude surveys, a classroom observation protocol, and an exemplar for using Rasch Analysis to analyze these tools. Quantitative reasoning does not have a home within any of the STEM subject areas, not even mathematics. There is a need for research on the teaching and learning of quantitative reasoning in STEM. How do we meet the challenge of sustaining interdisciplinary quantitative reasoning across subject matter silos which constitute today's schools?

References

Bartosh, O. (2003). *Environmental education: Improving student achievement.* Master Thesis, Evergreen State College.

Briggs, W. L. (2004). What mathematics should all college students know? In R. Gillman (Ed.), *Current practices in quantitative literacy, 17-19*. Washington, D.C.: Mathematical Association of America.

Corcoran, T., Mosher, F., & Rogat, A. (2009). *Learning progressions in science: An evidence-based approach to reform*. Philadelphia, PA: Consortium Policy Research in Education.

Dingman, S. W., & Madison, B. L. (2010). Quantitative reasoning in the contemporary world, 1: The course and its challenges. *Numeracy, 3*(2): Article 4. https://doi.org/10.5038/1936-4660.3.2.4.

Duschl, R. A., Schweingruber, H. A., & Shouse, A. W. (2007). *Taking science to school: Learning and teaching science in grades K-8*. Washington, D.C.: National Academies Press.

Elrod, S. (2014). Quantitative reasoning: The next "across the curriculum" movement. *Peer Review, 16*(3), 4–8.

Holland, J. H. (1992). Complex adaptive systems. *Daedalus*, 17–30.

Lehrer, R., Schauble, L., Carpenter, S., & Penner, D. (2000). The interrelated development of inscriptions and conceptual understanding. In P. Cobb, E. Yackel, & K. McClain (Eds.), *Symbolizing and communicating in mathematics classrooms: Perspectives on discourse, tools, and instructional design* (pp. 325–360). Mahwah, NJ: Lawrence Erlbaum Associates.

Lieberman, G. A., & Hoody, L. L. (1998). *Closing the achievement gap: Using the environment as an integrating context for learning*. Results of a Nationwide Study.

Madison, B. L. (2003). Articulation and quantitative literacy: A view from inside mathematics. In B. L. Madison & L. A. Steen (Eds.), *Why numeracy matters for schools and colleges* (pp. 153–164). Mathematical Association of America.

Madison, B. L., & Steen, L. A. (Eds.). (2003). *Quantitative literacy: Why numeracy matters for schools and colleges*. Princeton, NJ: National Council on Education and the Disciplines.

Mayes, R., & Koballa, T. R., Jr. (2012). Exploring the science framework. *Science and Children, 50*(4), 8–15.

Mayes, R., & Myers, J. (2014). *Quantitative reasoning in the context of energy and environment: Modelling problems in the real world*. Rotterdam: Sense Publishers.

Mayes, R. L., Forrester, J. H., Christus, J. S., Peterson, F., & Walker, R. (2014a). Quantitative reasoning learning progression: The matrix. *Numeracy, 7*(2): Article 5, 1–20.

Mayes, R. L., Forrester, J. H., Christus, J. S., Peterson, F., Bonilla, R., & Yestness, N. (2014b). Quantitative reasoning in environmental science: A learning progression. *International Journal of Science Education, 36*(4), 635–658.

Mayes, R. L., Peterson, F. I., & Bonilla, R. (2012). *Quantitative reasoning in context.* Retrieved from: www.uwyo.edu/wisdome/_files/documents/qrincontext_mayespeterson.pdf.

Mayes, R. L, Peterson, F., & Bonilla, R. (2013). Quantitative reasoning learning progressions for environmental science: Developing a framework. *Numeracy, 6*(1): Article 4.

MUSE. (2015). *Modelling for understanding in science education*. Retrieved from http://ncisla.wceruw.org/muse/.

National Environmental Education and Training Foundation (NEETF). (2000). *Environment-based education: Creating high performance schools and students.* Retrieved from http://www.neetf.org/pubs/NEETF8400.pdf.

National Governors Association Center for Best Practices, Council of Chief State School Officers. (2010). *Common core state standards mathematics.* Washington D.C.: NGAC, CCSSO.

NGSS Lead States. (2013). *Next generation science standards: For states, by states*. Washington, D.C.: The National Academies Press.

PLTW Project Lead the Way. (2017). *PLTW K-12 pathways*. Retrieved from https://www.pltw.org/our-programs/pltw-engineering.

Ramful, A., & Ho, S. Y. (2015). Quantitative reasoning in problem solving. *Australian Primary Mathematics Classroom, 20*(1), 15–19.

Savery, J. R. (2006). Overview of problem-based learning: Definitions and distinctions. *Interdisciplinary Journal of Problem-based Learning, 1*(1), 3.

Schwarz, C. V., Reiser, B. J., Davis, E. A., Kenyon, L., Achér, A., Fortus, D., et al. (2009). Developing a learning progression for scientific modeling: Making scientific modelling accessible and meaningful for learners. *Journal of Research in Science Teaching, 46*(6), 632–654.
Steen, L. A. (2001). *Mathematics and democracy: The case for quantitative literacy.* Princeton, NJ: The Woodrow Wilson National Fellowship Foundation, The National Council on Education and the Disciplines.
Strobel, J., & Van Barneveld, A. (2009). When is PBL more effective? A meta-synthesis of meta-analyses comparing PBL to conventional classrooms. *Interdisciplinary Journal of Problem-based Learning, 3*(1), 4.
Thompson, P. W. (1993). Quantitative reasoning, complexity, and additive structures. *Educational Studies in Mathematics, 25*(3), 165–208.
Thompson, P. W. (2011). Quantitative reasoning and mathematical modeling. In S. Chamberlin, L. Hatfield, & S. Belbase (Eds.), *New perspectives and directions for collaborative research in mathematics education* (pp. 33–37). Laramie, WY: University of Wyoming.
Thompson, P. W., & Saldanha, L. (2000). Conceptual issues in understanding sampling distributions and margins of error. In M. L. Fernández (Ed.), *Proceedings of the Twenty-second Annual Meeting of the International Group for the Psychology of Mathematics Education.* Tucson, Arizona.
U. S. Department of Education. (2015). *Science, technology, engineering and math: Education for global leadership.* Retrieved from https://www.ed.gov/stem.
Wilensky, U., & Resnick, M. (1999). Thinking in levels: A dynamic systems perspective to making sense of the world. *Journal of Science Education and Technology, 8*(1), 3–19.
Williams, D. C., Pedersen, S., & Liu, M. (1998). An evaluation of the use of problem-based learning software by middle school students. *Journal of Universal Computer Science, 4*(4), 466–483.
Wing, J. M. (2006). Computational thinking. *Communications of the ACM, 49*(3), 33–35. https://doi.org/10.1145/1118178.1118215.

Chapter 9
Modelling and Programming of Digital Video: A Source for the Integration of Mathematics, Engineering, and Technology

Carlos A. LópezLeiva, Marios S. Pattichis and Sylvia Celedón-Pattichis

Abstract Whilst Science, Technology, Engineering and Mathematics (STEM) interdisciplinary teaching and learning in the USA K-12 education still needs greater promotion, middle school students demonstrated that they can, using low-cost, single board computers that promote the teaching of computer science (in this case Raspberry Pis), successfully engage with computer programming of digital images and videos. The context for these students' engagement was the Advancing Out-of-School Learning in Mathematics and Engineering (AOLME) Project. This chapter describes how the processes of design, model, and implement, supported 40 Latinx middle school students' development of computational thinking in an out-of-school setting, and how these processes promoted the genuine integrated practice and learning of technology, engineering, and mathematical concepts.

Keywords Computer programming · Interdisciplinary modelling · STEM education · Out-of-school environment · Middle school students · Image and video representations

9.1 Introduction

Instruction of Kindergarten to Year 12 (K-12) subjects, especially mathematics, is often presented in isolation. Similarly, engineering, as a subject, is rarely integrated with science and mathematics; in fact, it is predominantly absent in the compulsory school curriculum and rarely an option as an elective (Celedón-Pattichis, LópezLeiva, Pattichis, & Llamocca, 2013). As a result, K-12 students get very little to no exposure to engineering (Katehi, Pearson, & Feder, 2009). It is especially the case of many middle (Grades 6–8 in the USA) school students, who often have limited access to experiences and information about interdisciplinary knowledge, such as science, technology, engineering, and mathematics (STEM) fields. Unfortunately, evidence reveals that this omission often translates into: student misconceptions of what STEM

C. A. LópezLeiva (✉) · M. S. Pattichis · S. Celedón-Pattichis
University of New Mexico, Albuquerque, NM 87131, USA
e-mail: callopez@unm.edu

B. Doig et al. (eds.), *Interdisciplinary Mathematics Education*,
ICME-13 Monographs, https://doi.org/10.1007/978-3-030-11066-6_9

fields are about; the exclusion of STEM careers from students' future goals (Mooney & Laubach, 2002); and the view of school subjects as scattered, artificial, knowledge, with little relevance to their lives (Reeves, 2011). Current statistics in the USA reveal misconceptions of STEM fields, for example, that learning technology means knowing basic computing and internet browsing, or, that STEM fields are too difficult, boring, or exclusionary, especially to students from minority backgrounds (Hossain & Robinson, 2012). Not surprisingly, the under-representation of ethnic groups and women in STEM fields in the USA is evident (Dowd, Malcom, & Macias, 2010; Landivar, 2013; Syed & Chemers, 2011) and so is the low enrollment of high school students in STEM academics, as well as their decreasing interest of staying in school (Hossain & Robinson, 2012; Rogers, 2009). Alarmingly, while there is an increasing demand for computing graduates in the USA (Defense Advanced Research Projects Agency DARPA, 2010), there is a declining student enrollment in Computer Science degrees, with high dropout, and failure rates, during the first and second years in the programme (Beaubouef & Mason, 2005). Correspondingly, Latinx high-school and college students in the USA have faced limited success in, and access to STEM fields, the highest dropout rates, and the lowest graduation percentages (Chapa & De La Rosa, 2006; Ortiz, Valerio, & Lopez, 2012).

Consequently, there is a strong need for the student population in general, specifically for minority ethnic groups, and women, to develop a scientifically literate citizenry, and inspiration to pursue STEM fields. The integration of science and engineering concepts and practices with mathematics need to become more evident to students at schools (Shaughnessy, 2012). Current educational frameworks, like the Common Core State Standards for Mathematics, CCSS-M (National Governors Association for Best Practices & Council of Chief State School Officers, 2010), and the Next Generation Science Standards (NGSS Lead States, 2013) have recommended such integration. Interdisciplinary approaches reach beyond by providing students with learning experiences that, not only include tasks and problems solved through subject integration, but also that depict more genuine applications of the topics learned. In this chapter, we describe the participation of USA middle school, predominantly Latinx students, in the Advancing Out-of-School Learning in Mathematics and Engineering (AOLME) Project, and how, through a model-eliciting activity (MEA) (Lesh, Hoover, Hole, Kelly, & Post, 2000) of programming digital videos, students engaged in technology, engineering (specifically computer programming), and mathematics practices. Analysis of students' practices and reasoning, using low-cost, single board computers designed to promote teaching computer science, called Raspberry Pis, yielded a three-stage process (i.e., designing, modelling, and implementing) for the creation of such projects.

The design, model, and implement processes, describe students' reasoning through MEAs and how they developed their videos from original sketches, using paper and pencil, all the way through the programming and sharing of their final projects. Through design students generated the content, or the stories, characters, and images to be included in their digital video. During modelling with mathematics, students mathematised the video images using two-dimensional array and their colours by using binary and hexadecimal numbers. In implementing, they coded the mathematised information of the video. The coding process involved the use of an actual programming language (Python in a Linux environment) to program and represent the images of the digital videos. In this chapter we will describe, through samples of students' work on MEAs, how the designing, modelling with mathematics, and implementing processes, evolved. We argue that these processes are integrative for the teaching and learning of technology, engineering (computer programming), and mathematics, especially when such MEAs include students' creativity, interests, and current knowledge.

9.2 Methods

The Advancing Out-of-school Learning in Mathematics and Engineering (AOLME) project was initially developed during a three-year period, with the initial participation of a total of 50 middle school (grades 6–8) students, predominantly Latinx, who attended one of the four 10–12 week programmes. In this manuscript, we analyse the processes that took place during the creation of the final projects. The final projects were set up through a MEA framework. Given the quality of student engagement documented through our participant observation, we chose to focus exclusively during the three sessions where students worked in the final project rather than the other sessions where students learned about and practice the curricular content. Thus, the final projects represent an application of what students learned in the project to program collectively a colour video from the pixel level. The data corpus for this study included the three videotaped sessions of students groupwork during the four AOLME programmes. In each programme students worked in teams, with total of about 5 teams per programme. Other data also included student interviews and facilitators' field notes.

Based on a participatory, situated engineering learning perspective (Johri & Olds, 2011), we analyzed the data knowing that the development of engineering and mathematical identities is parallel to the learning process (Litzinger, Hadgraft, Lattuca, & Newstetter, 2011; Martin, 2006) and that student learning is relational (Domínguez, LópezLeiva, & Khisty, 2014). A saturated comparative, and contrastive analytical data process (Miles & Huberman, 1994) yielded that the creation of video representations resulted from a three-fold process (i.e., design, modelling, and implementation) that simultaneously combined students' mathematical and engineering perspectives with students' current resources and understandings, since students drew from own interests and collectively-generated ideas to creatively develop and program videos using Python code. To describe this process, we selected the work of a specific team of students to present a storyline of how collectively students engaged in the design, modelling, and implementation of videos.

9.3 The AOLME Project

The Advancing Out-of-school Learning in Mathematics and Engineering project, AOLME sets out to motivate middle school students, predominantly Latinx, in urban and rural settings, to engage in more meaningful experiences in Technology, Engineering, and Mathematics fields, and to increase their interest to pursue related careers. AOLME is currently (see aolme.unm.edu) implemented in two schools (rural and urban) in the Southwest of the USA, with a total of 40 students yearly. Through an integrated mathematics and computer engineering curriculum, AOLME focuses on authentic learning experiences about the work of computer engineers, such as computer programming, engineering design, and mathematical concepts. Student learning and attitudes toward and interests in Technology, Engineering, Mathematics, and computing careers encompass the focus of the research project. AOLME's ultimate goal is to first finalize the curriculum, and then adapt it to be implemented as an elective course, during regular school time.

The AOLME project was developed through the recommendations of prior research (Celedón-Pattichis et al., 2013) that suggest interdisciplinary approaches to support greater learning opportunities. For example, in bilingual education the integration of content areas into a lesson, or unit, is considered as a meaningful approach, as students get to experience and learn the target concepts in a contextualised and interconnected environment that provides a set of generative ideas that support the development of conceptual understandings, discourses, terminology, practices, and knowledge pertinent to and needed in that context (Pérez & Torres-Guzmán, 2002). This integration must extend also to the interdisciplinary participation of experts from different fields, such as, educators and engineers. Unlike AOLME, prior efforts attempting to include engineering, in earlier grades, have mostly been done by people who have been trained in the engineering field and not by people with specific training and research in education (Johri & Olds, 2011; Litzinger, Hadgraft, Lattuca, & Newstetter, 2011).

The mathematics and engineering curriculum prepares students through a twelve-session (2–3 h each session) programme to design and implement digital image and video representations. The curriculum includes pencil-and-paper, modelling, and computer-based tasks. Topics covered introduce students to building a basic computer system; using co-ordinate systems to represent images and videos; and moving across number systems (e.g., binary and hexadecimal) to represent, model, and program their images and videos. The final project requires students to develop their own designed images and video. Initially, AOLME used Matlab, since it represents the dominant computing environment for applied mathematics and numerical analysis (Strang, 2007). The high cost of Matlab led us to select Python as the free and open-source alternative that is also very easy to install and use for scientific computing applications (Jones, Oliphant, & Peterson, 2014). Python is a popular programming language and is freely and openly available on all computing platforms. We also selected the Raspberry Pi computer, since it is a low-cost, single board computer that supports Python, Linux languages, and Mathematica software (e.g., Canelake, 2011).

Furthermore, AOLME developed Python libraries of image and video processing functions that have allowed us to provide the functionality needed for the project. The video programming process is also linked to a modelling process that we describe next.

9.4 Mathematical Modelling

Mathematical modelling presents a context with great potential for the integration of technology, engineering, and mathematics. This approach aims to motivate schools, teachers, and students to engage in mathematics modelling that integrates mathematics with purposeful applications. In the case of AOLME, mathematical modelling is integrated with engineering, through the development of digital images and video. On the one hand, digital image and video representations provide a learning context in which engineering and mathematical concepts blend naturally, and where students are challenged to learn the underlying mathematical structure and computer programming concepts, required for practical implementation. On the other hand, AOLME students have demonstrated that they are highly motivated to participate in activities that include the development of digital images, and videos, that they can also share with their friends. Thus, our approach represents a balanced context for motivating, and yet challenging students to engage in the practice of modelling with mathematics.

Previous studies have described the benefits of modelling with mathematics, which represents relationships found in everyday life in the mathematical system by using mathematical conventions and symbols (Lesh & Doerr, 2003). Thus, Modelling is a way for students to re-invent or "formalise their informal understandings and intuitions" (Gravemeijer, Cobb, Bowers, & Whitenack, 2000, p. 237) and rehearse mathematical concepts (Lesh & Doerr, 2003). Modelling helps students learn mathematics since "learning it means doing mathematics, in which solving real-life problems is essential" (Gravemeijer, 1997, p. 332). Then, modelling is a mathematical re-discovery process of patterns and rules in authentic situations. It represents a high level of cognitive demand for students of doing mathematics (Stein, Smith, Henningsen, & Silver, 2000). The modelling context matters to students since if they know the context to be modelled, the mathematical descriptions will flow more easily (Gravemeijer et al., 2000; Lesh & Doerr, 2003). In mathematics education in the USA, the Common Core State Standards for Mathematics (CCSS-M) initiative, defines Modelling as "the process of choosing and using appropriate mathematics and statistics to analy[s]e empirical situations, to understand them better, and to improve decisions" (National Governors Association for Best Practices & Council of Chief State School Officers, 2010, p. 72).

9.4.1 *Our Model-Eliciting Activities (MEAs) Framework*

Drawing on Lesh, Hoover, Hole, Kelly, and Post (2000), we use the model-eliciting activities (MEAs) framework to refer to activities that encourage students to develop and test models. On the one hand, MEAs help students understand that mathematics is more about doing, than just seeing. On the other hand, MEAs also support the assessment of students' higher-order understandings of computational thinking by developing models that reveal their interpretation of, thinking about, and mathematisation of images and videos. Thus, the model-eliciting activity framework supports AOLME work by promoting student learning and engagement in Technology, Engineering, and Mathematics practices, and by documenting how students reveal and reflect on their thinking through the modelling process. MEAs have informed our understanding of student' evolving learning and knowledge by being able to link the final products and intermediate solutions of the tasks with what the students documented and thought during the creation of these products. MEAs are developed based on the following principles:

1. Model Construction
2. Reality
3. Self-Assessment
4. Construct Documentation
5. Construct Shareability & Reusability, and
6. Effective Prototype (Chamberlin & Moon, 2005; Lesh, Amit, & Schorr, 1997, 2000).

AOLME includes only one model-eliciting activity at each curricular level. Each accounts for about 30% of the time that students spend during a specific level of the AOLME curriculum. For example, during Level 1 (included in this chapter), students design, and program, a digital video (black and white or colour from the pixel level), collectively. The process starts with the generation of a story for the video that represents the group's identity and interests. Then, they model with mathematics the images or frames in the video and translate these models into an intermediate code that helps them to program and display the envisioned original story (see later Figures of students' work). The model-eliciting activity is open-ended to support students' input regarding their prior skills, knowledge, and needs, through which they design and develop a product that reflects their personal interests. Throughout, students document these processes using different formats (e.g., paper and pencil, digital files, or scripts of their coding) that describe how their ideas and plans evolved over time. As a result, the whole process asks students to develop collectively explanations and constructions by "repeatedly revealing, testing, and refining or extending their ways of thinking" (Lesh et al., 2000, p. 597). MEAs simultaneously integrate student participation in, and learning of, computational thinking, mathematics, art, and computer programming. Because students constantly revise and reason about their models, they actively and genuinely use discourse related to computational thinking, computer programming, and related mathematical concepts. Because "students develop mathematics [Technology, Computer, and Engineering] concepts as they use them discursively to construe meaning" (Schleppegrell, 2007, p. 148), students engage in and develop interdisciplinary discursive (i.e., thinking, talking, and doing) practices through their work in projects, or MEAs.

9.5 Findings

The designing, modelling, and implementation phases comprise a set of non-linear processes that are continuously linked. These connections are especially evident when students debug the program interactively based on the output of their work. This portrayal informs them about alignment with their design, modelling with mathematics, and the implementation. Thus, student-generated videos, or their final projects, blend together the designing, modelling, and implementation processes in the development and refinement of student goals.

9.5.1 The Process of Designing

Designing relates to the student generation of content and ideas of the stories and characters that they include in their digital images and video. The design process emerges from students' ideas as sources for modelling with mathematics and supports their authoring of visual narratives. This process engages students in trying out new identities (Gee, 2003) through engagement in new practices. It allows them to take ownership of the modelling process and creation of the video, not only because they create it, but also because the source of the project itself includes their own ideas. We believe that this situation represents a Realistic Mathematics Education approach, because we witnessed how students' attitudes during Modelling and programming "went from not caring much to actually being genuinely interested in the topic, especially as we started working with video" (Facilitator's Fieldnotes, Summer 2013). Also, students "were proud of what they did and they enjoyed showing others and watching it themselves. They were a lot more interested to find out how it works" (Facilitator's Fieldnotes, Spring 2014). We believe that student-centred video creation is a source of student real interest in mathematics and engineering applications. Videos represent students' own ideas and self-expressions that are portrayed through digital image and video representations. As a result, the design of videos as an MEA, supported extensive reasoning, collaborative, and spontaneous creative work.

9.5.1.1 Extensive Collaborative Work

Since the video was to represent a team, the process of conceptualising the video was often a collaborative process, where the brainstorming of own ideas and shared interests and experiences in the team, were the sources of creativity and storylines. Teams determined their team names at the beginning of the programme, and, as they shared being part of the same team, a shared collective identity around that name emerged and influenced the creative process. In the case of the Eagles team, students used a picture of a frog during the session on colour images. This frog was seen as cute and funny. The team brainstormed about the video topic, and the discussion evolved as follows:

Dora:	We can probably do a car, like a frog jumping.
Jonathan:	I like the frog [searching in mobile phone]
Dora:	[continues story] …into a hood, changing its colours like a chameleon and then jumps into the car and then it hits the car and then it jumps to another car and it gets into a wall and the car explodes! [expands arms].
Alejandro:	Yeah, we have to make it like really …
Juan:	But, the frog dies! [whines and smiles]

The interaction above provides evidence of a participatory and shared process of generating the storyline of the video. While some aspects changed, most of these topics remained throughout the completion of the project. After agreeing with the story, the division of labour to complete the work for the video was as shared as the development of the storyline. The interaction below illustrates the negotiation of this distribution of character designs.

Dora:	Let me see how the frog is (i.e., looks) [says to Jonathan].
Juan:	I'll do the car.
Alejandro:	So here is the frog [shows grid with frog to Juan].
Juan:	[nods]
Alejandro:	We need the jumping frog, then we need the car, and I think that's it.
Alejandro:	[to Dora] This should be yellow [points to frog's mouth], this should be black [frog's eye], is that right?

Team members consulted each other throughout the image design process, on the colours and the tasks to complete the video story. The openness of final project as a MEA helped teams capitalise on members' diverse interests and skills and collaborate as a result. When issues arose in the design process, the collective brainstorming, collective support, and creativity increased as described next.

9.5.1.2 Spontaneous Work and Creativity

The design process promoted shared leadership and team-work. The energy around the teams was characterised by smiles, dedicated work, intense thinking processes, consultation, investigation, reasoning, communication, and creativity. Figure 9.1 depicts such an environment.

The design of the car for the frog encompasses this collective and creative process. While initially the idea of having the frog in a car was clear, what seemed unclear was the size and look of the car and the frog. Another issue, or constraint, related

Fig. 9.1 Eagles team working on the initial design of their video frames

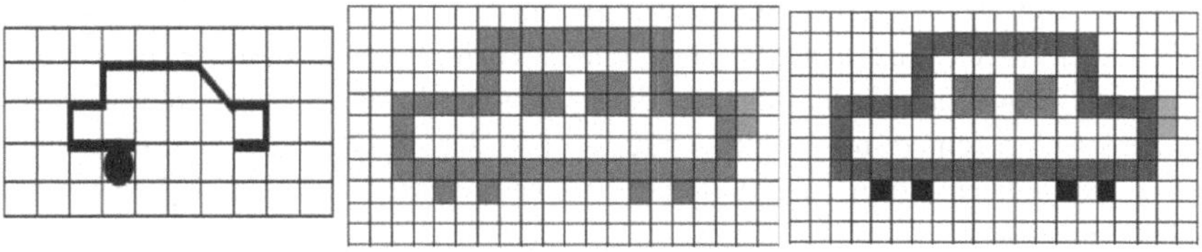

Fig. 9.2 Initial car designs

to these elements, was the size of the matrix, or number of pixels, for each video frame. Figure 9.2, depicts the evolving shape of the car. The picture on the left was the first car design that was changed since the wheels, the front side of the car, and the windshield did not include complete pixels. In fact, this image is just an outline image rather than a pixellated image. Then, the car in the middle came to be a design that aligned better to these constraints. Once the car aligned with the constraints of the design for pixels, Dora selected colours for the car (picture on right).

Nevertheless, the frog design presented a challenging task because it needed to be small enough in relation to frame size to allow other aspects of the story to be included, like the frog movement of jumping onto and driving the car, but still the image needed to include enough details to be identified as a picture of a frog. Thus, once the design of the frog was completed, it affected the design of the car. Figure 9.3 on the left depicts the final frog design on paper. The car picture in the middle shows how the car shape above was enlarged to fit the frog inside. The new design seemed inefficient in several ways as it only had 5 pixels of height and the frog was 6 pixels tall. Also, the frog needed to jump into the car, so the car design seemed limiting to the story and the frog movement. In collaboration with the team, Dora developed a new car design, a convertible (picture on the right). This design perfectly fit the story

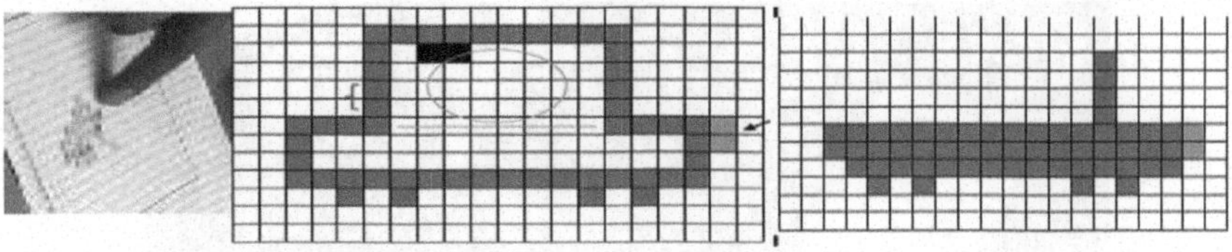

Fig. 9.3 Relationship between the frog and the car designs

and size of the frog. The collection of car designs shows a negotiated, collective, and creative process of the design of the most efficient image through the interplay of the team's story, goals, mathematical reasoning, and communication, and under the constraints of developing or creating pixellated images.

While the team distributed the design of elements (objects, frames, or sequence of images in the story) among their members, these elements needed to come together in a cohesive way to align to the common vision and goal of the video. In this process, negotiation of differences, reciprocal support, measuring, shared reasoning, and creativity, were spontaneous processes in the team's effort of design. Students perceived their videos as a real activity that demanded applications of mathematical and engineering concepts. Under this lens, the sequences of frames, or digital images, of the videos represented meaningful sources for student engagement and the promotion of computer programming and mathematical practices such as modelling with mathematics.

9.5.2 The Process of Modelling with Mathematics

While designing is also linked to measuring the shape and size of objects given a coordinate plane, the second process, to model with mathematics, refers to a digital modelling process, or the binarisation (black and white- or 0 and 1, or hexadecimal representations of colours) of the images. Using two-dimensional array representations or a coordinate plane, students generated images, by using grid paper and pencil. Now, the next step is to mathematise, or model, these images from the pixel level by indexing specific location and colour of the squares or pixels, on the coordinate plane. The images in each frame were to be transformed throughout the video, so that each frame would continue, visually and conceptually, the storyline that students envisioned. The transcript of the interaction below presents how the team continued working together to organise the order of the frames, as well as to mathematise, and index the colours, by using hexadecimal numbers. Specifically, Juan was working with the facilitator on the frame where the frog crashes the car against a wall.

Dora: OK, this is one, two, this one is three, this one is four [writes a number to order the frames].

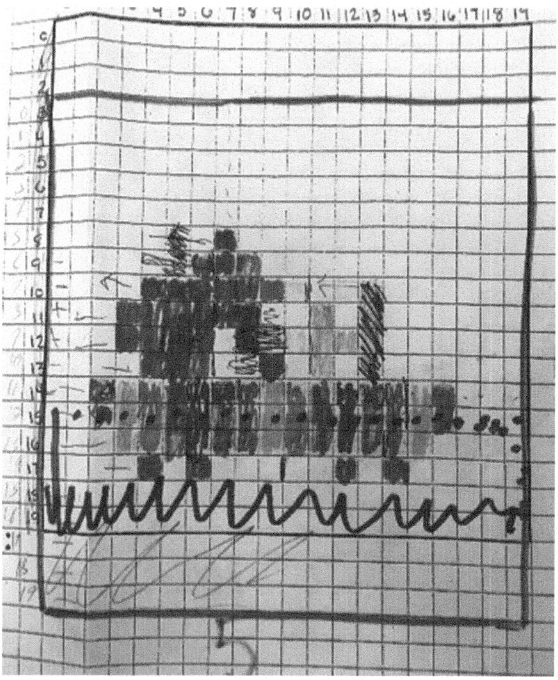

Fig. 9.4 Mathematising the frog on multicoloured car

Juan:	The variables we'll use for the colours are: G = green, R = red, GR = grey.
Alejandro:	What colour is for the brick?
Juan:	Red? … the orange is F, F, A, D.
Alejandro:	A, D
Juan:	Zero, zero
Alejandro:	Zero, zero, ok.
Juan:	Brown is ninety-nine

As Fig. 9.4, shows, students numbered each frame with the goal of identifying specific pixels. This coordinate system would ease the location of the pixels to enter the previously designed images. In this example, one can see how at this point, the car was conceived as multicoloured, a fact that was changed later on into a monochrome car to ease the shifting colours features of the frog. For the indexing of the location of the pixels in the car, one can see how students used black dots to mark the exact location. They also used arrows to mark the number of pixels that the frog was away from the border of the frame. Notice how the frame was lowered to centre the frog.

To introduce and conclude the video, the team also designed an image of an eagle to show ownership of the video. The features of the design of the eagle became more evident to the group through the Modelling with Mathematics process, as students noticed that the wings and body of the eagle were symmetrical. This discovery helped them simplify the work by only indexing a half of the eagle's co-ordinates.

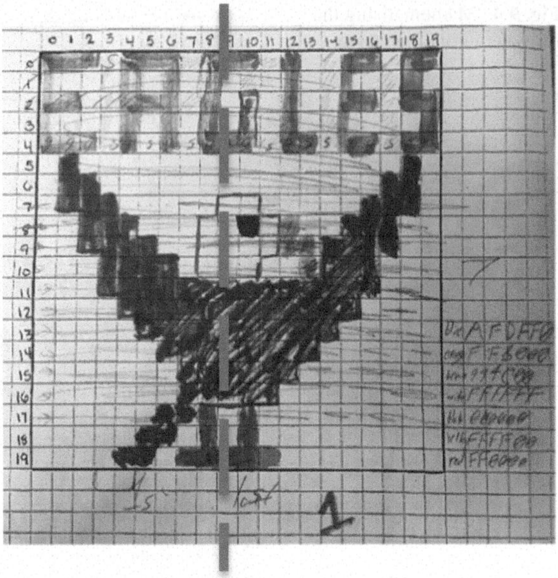

Fig. 9.5 Frame one of the Eagle team's video: a symmetric eagle

Figure 9.5 displays the line of symmetry that students identified. At the same time, students listed the colours with the corresponding hexadecimal numbers at the lower right corner of Fig. 9.5 to ease the entry data for this video frame. Students simply entered half the codes for the wings, legs, and blank spaces around it and then copied it. Then, students added data details that varied from the symmetry pattern to complete full eagle image. As a result of student investment through the development of their own ideas of images and expression of who they were as a team, students felt encouraged not only to complete the look of the project in the best way possible, but also to notice and strive for mathematical and computer programming syntactic accuracy, so their project could be what they had envisioned. These aspects became evident in the implementation process.

9.5.3 The Process of Implementing

The last process, to implement, refers to developing the coding of the representations of the digital images, and video, by using two-dimensional arrays. This process involves a more direct connection to computer programming by programming with Python to represent digital images and video. In terms of modelling with mathematics, this process includes a mathematical component, coding for colours, which is directly integrated in computer science and engineering. Figure 9.6 depicts how students programmed the frog to shift colours as a chameleon, which took a complicated pathway. In consultation with an AOLME program facilitator, the Eagles team used

```
From AOLMEfun import
Import image
nrows = 20; ncols = 20

G1= "00FF00"
GY= "009900"
R= "FF0000"
O= "FFAD00"
B= "053CF0"
BL= "000000"
S= "AFDFF0"
BN= "994C00"
W= "FFFFFF"
Y= "FFFF00"
```

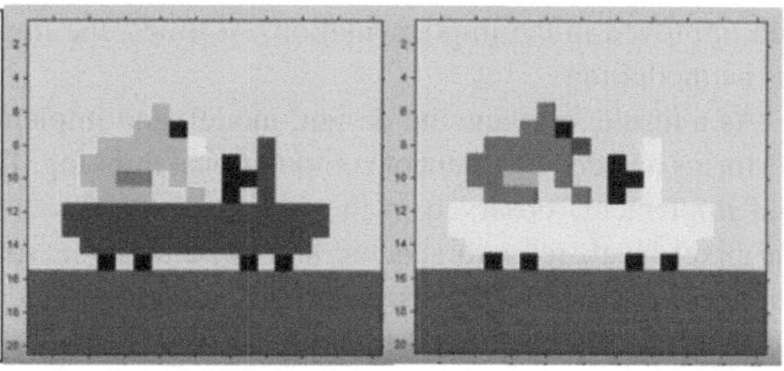

Fig. 9.6 Coding the chameleon frog

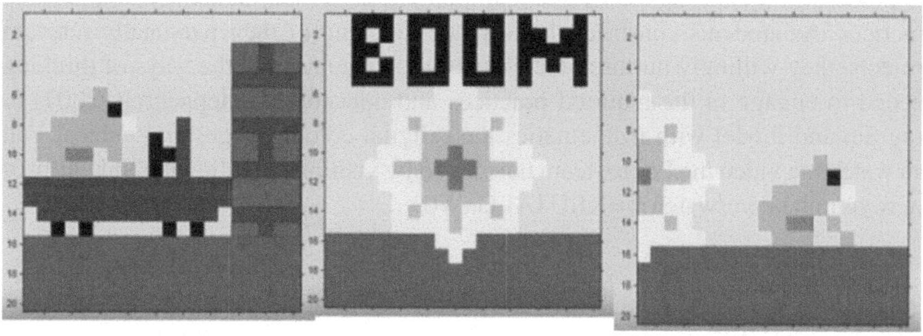

Fig. 9.7 Design of the crash through implementation

variables to ease the code entering process for each of the colours in the frames, especially for frames wherein the frog was at the same location, but needed different colours. Students created a variable for each colour to avoid repeating the color codes in the program. The image on the left describes the Python codes the team used to define each variable. The images in the middle, and on the right side, depict the shifting colours of the frog. For this to happen, the team simply copied the same Python color representation from the previous frame and, to change the colours, the team simply redefined the colour variable definitions each frame. In this way, the team saved time and effort and, at the same time, this process reveals the level of sophisticated understanding that the team managed to achieve. Additionally, this example describes the relationship and fluid movement across the design, model, and implement processes.

Another aspect that became important to the team was the frame-rate or speed of displaying the video, so that the shifting of colours and the crash would look as desired. The team decided, collectively, on this feature through trial of several speeds until consensus was reached. As Fig. 9.7 portrays, the Eagles team broke down their story into frames that told the story in the most efficient way. The team increased cohesion of the frames by adding a common background across frames (brown ground and blue sky). Such features became more evident as the team ran, or implemented, their video. This feedback process describes the interconnected functions of the design, model, and implement processes. The design can be changed and improved in the implementation. At times, the implementation revealed issues of the modelling.

As a result, we view the design, model, and implement processes related to the nurturing and development of computational thinking. The processes (design, model, and implement) observed in the MEA for the development of a digital video at the pixel level, refer to specific stages of thought, social support, and emotional engagement. Together, these elements supported student computational thinking. Further, this computational thinking seems to make sense to the students because the whole process arises from the students' ideas, interests, needs, and goals. The language of computers and mathematics become tools to achieve these goals. In the process, students engage in, and learn, engineering and mathematical discourse practices. As students constructed and made meaning of their personally-relevant projects, they willingly nurtured their use of the language and the ways of thinking needed to engage in the required practices and concepts (Schleppegrell, 2007) to program and model with mathematics their digital colour images and videos. You can watch the video the Eagles team developed by visiting the following link: https://www.youtube.com/watch?v=XJlUTALGaW0.

9.6 Discussion

The design, model, and implement processes, during a MEA, provide a promising integrative combination for the teaching and learning of mathematics and engineering, and computational thinking. Computational thinking processes emerge through a synergy between engineering design, mathematical modelling, and computer programming and student input. Computational thinking (CT) is defined as:

> CT is an approach to solving problems in a way that can be implemented with a computer. Students become not merely tool users but tool builders. They use a set of concepts, such as abstraction, recursion, and iteration, to process and analyse data, and to create real and virtual artifacts [sic]. CT is a problem-solving methodology that can be automated and transferred and applied across subjects (Barr & Stephenson, 2011, p. 52).

In relation to the design, model, and implement processes, computational thinking becomes especially significant when the process includes MEAs, or projects, that address students' creativity, interests, and current knowledge, such as video and image representations. This combination promotes the development of learning and ownership of technology, engineering, and mathematical practices and related identities. In this way, a genuine and meaningful engagement with the practices also nurtures a new lens of seeing oneself—new identities, possibilities of being—as creators, programmers of videos, modellers, and users of mathematics who translate their ideas and stories into numbers and digital representations. As a result, we learned that all middle school Latinx students in the programme successfully engaged in computational thinking and STEM practices. Further, images and video provided a real, and meaningful, context for students to learn about, and improve, programming skills. Also, collaboration in the teams seemed to be nurtured when student participation took place in an activity that was interesting to all, and in which the responsibility and authorship of the project were distributed among all participants. Finally, the discrete work that took place in each process (design, model, or implement) and the linear connections to the prior, or the next, process evolved when the understanding of a practice was clear and shared. Complex connections among the processes emerged through questions and interests that turned into a collective problem posing, or problem-solving task, that required the activation and interaction of all the processes. Greater analysis of the link between student participation, engagement in STEM practices, and identity development, from the perspective of the students, would be illuminating. Exploration of collective problem solving through the complex interaction of the design, model, and implement processes provide further insight into mathematics, computer programming, and student thinking and learning.

Our society is evolving at a fast pace, where technological and cultural changes are evident. Under such demands, a transformative pedagogy is needed to support human development that grants greater access to current needs and re-humanizes mathematics (Gutiérrez, 2018). This pedagogy acknowledges power distribution, and accordingly, emancipates students from relegated positions through access to knowledges, opportunities, and new channels, that provide paths to navigate and engage meaningfully in technological and cultural forces that aim at positive impact on their communities (Code for All, n.d.; Cope & Kalantzis, 2009; Johri & Olds, 2011; Kalantzis, 2006; Kalantzis & Cope, 2005) and selves. Thus, we argue for a pedagogy that sustains students' cultural and linguistic competence of their communities, while also offering access to dominant cultural competence (Paris, 2012). This transformative, sustaining, and responsive STEM pedagogy—while sensible to students' cultural and linguistic backgrounds—also envisions and supports students as creators and producers, rather than simply consumers of meaning and technological knowledge and artefacts.

Although the Next Generation Science Standards describe computer programming, or coding as part of the curriculum in the U.S. (NGSS, 2013), through our experiences and literature, we have corroborated and witnessed the limited access that middle school students from minoritized groups have during regular school time and as part of the curriculum. Instead, school districts often reinforce remedial curricula and programs for schools with this population. Thus, these students end up having little to no access to knowledge and practices such as computer programming (Chapa & De La Rosa, 2006; Katehi et al., 2009; Mooney & Laubach, 2002) and the related mathematics. Elective courses and after-school programs are main sources of this limited access. Additionally, such programs are often led by experts outside of education fields (Johri & Olds, 2011; Litzinger et al., 2011).

In contrast, as described above, we have also witnessed how Latinx middle school students can engage successfully in computer programming and related mathematics practices. Through the design, model, and implement processes, we established how students engaged with computer programming and computational thinking. In fact, programming is described as a decomposition process (Szlávi & Zsakó, 2017). Students, who self-selected to be in AOLME, solved the problem through the complex process of designing and programming images and videos through abstract thinking by modelling, and de- and re-composing the images (digital and on paper) through the analysis and development of patterns, recursion, symbolic representation, and algorithmic thinking (Chamberlin & Moon, 2005; Lesh et al., 1997, 2000; Lesh & Zawojewski, 2007; Psycharis & Kallia, 2017; Wing, 2006) to implement their video projects.

Moreover, these projects were linked to student ideas and topics that were meaningful to them, so through programming they expressed human thinking and likes (Pea, Kurland, & Hawkins, 1985). Then, while these practices were new to students, they appropriated these practices by using their home language, collaborating in teams with one another and their facilitators, and creating videos with content meaningful to themselves, their peers, and families. Simultaneously, students gained knowledge of practices that the facilitators—mostly engineering college students—identified as non-existent in their K-12 experience. Thus, AOLME experiences provided students with opportunities to see what they can do, as well as an improved understanding of what they need to know to go to college, especially for studying Engineering and Computer Science, where programming is essential. Given current technological demands in society, goals that students described for their future, and student opportunities to access knowledge that addresses both demands, there is growing evidence of a transforming, sustaining, and responsive STEM pedagogy in the AOLME approach.

Furthermore, we see the process of understanding the mathematics involved in computer programming as an essential knowledge that does not only situate the knowledge (Johri & Olds 2011; Pea et al., 1985) within the practice of computer programming, but is also identified by the students as transformative in at-least two different ways. On the one hand, students believed the learning of binary and hexadecimal numbers, the definition of variables, and the use of the coordinate plane, to design images, as a challenging process because it was all-new to them. On the other hand, students also argued that while the new mathematical knowledge is not directly related to school mathematics, due to their view of the challenging nature of this new mathematics knowledge, students gained greater confidence in themselves, and their mathematical ability to do mathematics in the classroom. Hence, almost all of them reported performing better in school mathematics. Nevertheless, while it is important to learn the mathematics to design, model, and program images and videos, it would be just as important to use mathematical ideas and content to enhance and learn computer programming knowledge. The AOLME curriculum aims to include such an approach as part of learning to represent digital images and videos. Again, such an approach will provide greater evidence for a transformative, sustaining, and responsive STEM pedagogy.

Acknowledgements First, we want to thank the students and facilitators that have participated in AOLME. They are the reason and synergy of this project. Secondly, we also want to thank Daniel Llamocca, Neritza Diaz-Cruz, and Jessica Morales for their contributions to this study. Finally, we acknowledge the support by the National Science Foundation under NSF ITEST Award #: 1613637,

Broadening Participation of Latinx Students in Engineering Using an Integrated Mathematics, Engineering and Computing Curriculum in Authentic, Out-of-School Environments. Any opinions, findings, and conclusions or recommendations expressed in this material are those of the author(s) and do not necessarily reflect the views of the NSF.

References

Barr, V., & Stephenson, C. (2011). Bringing computational thinking to K-12: What is involved and what is the role of the computer science education community? *ACM Inroads, 2*(1), 48–54.

Beaubouef, T., & Mason, J. (2005). Why the high attrition rate for computer science students: Some thoughts and observations. *ACM SIGCSE Bulletin, 37*(2), 103–106.

Canelake, S. (2011). *A gentle introduction to programming using python*. http://ocw.mit.edu/courses/electrical-engineering-and-computer-science/6-189-a-gentle-introduction-to-programming-using-python-january-iap-2011/.

Celedón-Pattichis, S., LópezLeiva, C. A., Pattichis, M. S., & Llamocca, D. (2013). An interdisciplinary collaboration between computer engineering and mathematics/bilingual education to develop a curriculum for underrepresented middle school students. *Cultural Studies of Science Education, 8*(4), 873–887.

Chamberlin, S. A., & Moon, S. M. (2005). Model-eliciting activities as a tool to develop and identify creatively gifted mathematicians. *Journal of Secondary Gifted Education, 17*(1), 37–47.

Chapa, J., & De La Rosa, B. (2006). The problematic pipeline: Demographic trends and Latino participation in graduate science, technology, engineering, and mathematics programs. *Journal of Hispanic Higher Education, 5*(3), 203–221.

Code for All. (n. d.). *What we believe*. Retrievable from: https://codeforall.org.

Cope, B., & Kalantzis, M. (2009). "Multiliteracies": New literacies, new learning. *Pedagogies, 4*(3), 164–195.

Defense Advanced Research Projects Agency (DARPA). (2010). *Computer Science—Science, Technology, Engineering, and Mathematics (CS-STEM) Education Research Announcement (RA)*. DARPA-RA-10-03.

Domínguez, H., LópezLeiva, C. A., & Khisty, L. L. (2014). Relational engagement: Proportional reasoning with bilingual Latino/a students. *Educational Studies in Mathematics, 85*(1), 143–160.

Dowd, A. C., Malcom, L. E., & Macias, E. E. (2010). *Improving transfer access to stem bachelor's degrees at Hispanic serving institutions through America COMPETES Act*. Los Angeles, CA: University of Southern California.

Gee, J. P. (2003). *What video games have to teach us about learning and literacy*. New York, NY: Palmgrave-Macmillan.

Gravemeijer, K. (1997). Solving word problems: A case of modelling? *Learning and Instruction, 7*(4), 389–397.

Gravemeijer, K., Cobb, P., Bowers, J., & Whitenack, J. (2000). Symbolizing, modelling, and instructional design. In P. Cobb, E. Yackel, & K. McClain (Eds.), *Symbolising and communicating in mathematics classrooms: Perspectives on discourse, tools, and instructional design* (pp. 225–274). Mahwah, NJ: Lawrence Erlbaum Associates.

Gutiérrez, R. (2018). The need to rehumanize mathematics. In I. Goffney & R. Gutiérrez (Eds.), *Annual perspectives in mathematics education: Rehumanizing mathematics for black, indigenous, and latinx students* (pp. 1–10). Reston, VA: National Council of Teachers of Mathematics.

Hossain, M., & Robinson, M. G. (2012). How to motivate U.S. students to pursue STEM (Science, Technology, Engineering and Mathematics) careers. *US-China Education Review, A*(4), 442–451.

Jones, E., Oliphant, E., & Peterson, P. (2014). *SciPy: Open source scientific tools for python*. http://www.scipy.org/. Accessed 7 May 2018.

Johri, A., & Olds, B. M. (2011). Situated engineering learning: Bridging engineering education research and the learning sciences. *Journal of Engineering Education, 100*(1), 151–185.
Katehi, L., Pearson, G., & Feder, M. (2009). *Engineering in K-12 Education: Understanding the Status and Improving the Prospects*. Washington, D.C.: National Academy of Engineering and National Research Council.
Kalantzis, M. (2006). Changing subjectivities, new learning. *Pedagogies: An International Journal, 1*(1), 7–12.
Kalantzis, M., & Cope, B. (2005). *Learning by design*. Melbourne, Australia: Victorian Schools Innovation Commission.
Landivar, L. C. (2013). *Disparities in STEM employment by sex, race, and hispanic origin*. American Community Survey Reports: U.S. Census Bureau. https://www.census.gov/prod/2013pubs/acs-24.pdf.
Lesh, R., Amit, M., & Schorr, R. Y. (1997). Using "real-life" problems to prompt students to construct conceptual models for statistical reasoning. In I. Gal & J. B. Garfield (Eds.), *The assessment challenge in statistics education* (pp. 65–83). Amsterdam, Holland: IOS.
Lesh, R., & Doerr, H. M. (2003). *Beyond constructivism: Models and modelling perspectives on mathematics problem solving, learning, and teaching*. Mahwah, NJ: Lawrence Erlbaum.
Lesh, R., Hoover, M., Hole, B., Kelly, A., & Post, T. (2000). Principles for developing thought revealing activities for students and teachers. In A. E. Kelly & R. A. Lesh (Eds.), *Handbook of research design in mathematics and science education* (pp. 591–645). Mahwah, NJ: Lawrence Erlbaum.
Lesh, R., & Zawojewski, J. (2007). Problem-solving and modeling. In F. Lester (Ed.), *Second handbook of research on mathematics teaching and learning* (pp. 763–804). Reston, VA: NCTM.
Litzinger, T., Hadgraft, R., Lattuca, L., & Newstetter, W. (2011). Engineering education and the development of expertise. *Journal of Engineering Education, 100*(1), 123–150.
Martin, D. (2006). Mathematics learning and participation as racialised forms of experience: African American parents speak on the struggle for mathematics literacy. *Mathematical Thinking and Learning, 8*(3), 197–229.
Miles, M., & Huberman, A. M. (1994). *Qualitative data analysis* (2nd ed.). Thousand Oaks, CA: Sage Publications.
Mooney, M. A., & Laubach, T. A. (2002). Adventure engineering: A design centered, inquiry based approach to middle grade science and mathematics education. *Journal of Engineering Education, 91*(3), 309–318.
National Governors Association Center for Best Practices, Council of Chief State School Officers. (2010). *Common core state standards for mathematics*. Washington, D.C.: Authors.
NGSS Lead States. (2013). *Next generation science standards: For states, by states*. Washington, DC: Achieve Inc.
Ortiz, C. J., Valerio, M. A., & Lopez, K. (2012). Trends in Hispanic academic achievement: Where do we go from here? *Journal of Hispanic Higher Education, 11*(2), 136–148.
Paris, D. (2012). Culturally sustaining pedagogy: A needed change in stance, terminology, and practice. *Educational Researcher, 41*(3), 93–97.
Pea, R. D., Kurland, D. M., & Hawkins, J. (1985). Logo and the development of thinking skills. In M. Chen & W. Paisley (Eds.), *Children and microcomputers: Research on the newest medium* (pp. 193–317). Norwood: Ablex Publishing Corp.
Pérez, B., & Torres-Guzmán, M. E. (2002). *Learning in two worlds: An integrated Spanish/English biliteracy approach* (3rd ed.). Boston, MA: Allyn and Bacon.
Psycharis, S., & Kallia, M. (2017). The effects of computer programming on high school students' reasoning skills and mathematical self-efficacy and problem solving. *Instructional Science, 45*(5), 583–602.
Reeves, T. (2011). Can educational research be both rigorous and relevant? *Educational Designer, 1*(4), 1–24.

Rogers, S. (2009, September). *Rapid prototyping: A strategy to promote interest in STEM careers.* Paper Presented on US-Turkey Workshop on Rapid Technologies. Accessed 16 March 2015: http://iweb.tntech.edu/rrpl/rapidtech2009/rogers.pdf.

Schleppegrell, M. J. (2007). The linguistic challenges of mathematics teaching and learning: A research review. *Reading & Writing Quarterly, 23,* 139–159.

Shaughnessy, J. M. (2012). *STEM: An advocacy position, not a content area. NCTM Summing Up.* Accessed 7 May 2018 https://www.nctm.org/News-and-Calendar/Messages-from-the-President/Archive/J_-Michael-Shaughnessy/STEM_-An-Advocacy-Position,-Not-a-Content-Area/.

Stein, M. K., Smith, M. S., Henningsen, M. A., & Silver, E. A. (2000). *Implementing standards-based mathematics instruction: A casebook for professional development*. New York, NY: Teachers College Press.

Strang, G. (2007). *Computational science and engineering*. Wellesley: Wellesley-Cambridge.

Syed, M., & Chemers, M. M. (2011). Ethnic minorities and women in STEM: Casting a wide net to address a persistent social problem. *Journal of Social Issues, 67*(3), 435–441.

Szlávi, P., & Zsakó, L. (2017). The cognitive toolkit of programming—algorithmic abstraction, decomposition-superposition. *Acta Didactica Napocensia, 10*(4), 33–40.

Wing, J. M. (2006). Computational thinking. *Communications of the ACM, 49,* 33–35.

Part III
Case Studies in Inter-Disciplinarity: Mathematics as Tool and Mathematics as (Conscious) Generalisation

David Swanson

Chapter 10
Introduction

David Swanson

Abstract The five case study chapters in this section are introduced and contextualised in relation to previous case study work in Interdisciplinary Mathematics Education. Alongside the benefits which come from interdisciplinary work, a common theme which emerges from the case studies is that of a potential for mathematics to disappear, or to become a mere tool, within such activities. Appealing to Vygotsky's theory of scientific concepts, it is argued that there is a crucial role for generalisation within interdisciplinary mathematics, and that the connections within mathematics require attention alongside the connections between mathematics and real world experience if mathematics is to more fully benefit from, and bring benefit to, interdisciplinary work.

Keywords Mathematics · Interdisciplinarity · Vygotsky · Scientific concepts · Generalisation

10.1 Case Studies in Inter-disciplinarity

In a previous survey of the literature on interdisciplinary mathematics (Williams et al., 2016), I noted that the activities researched within existing case studies divided fairly evenly into 3 categories: (i) those where mathematics appeared in another subject, or vice versa; (ii) thematic integration, where the subjects including mathematics were directed toward a common theme or project whilst retaining their particular disciplinarity; and (iii) project or problem based activity, where the problem or project had become the central question and individual subjects had begun to dissolve. It was argued that these categorizations mapped loosely (and with exceptions) onto the common categorizations of mono-disciplinary, multi-disciplinary, and inter-disciplinary. The case studies in this section arguably follow this pattern, with examples where mathematics appears in another subject; where thematic integration occurs, with dis-

D. Swanson (✉)
Manchester Institute of Education, The University of Manchester, B4.10 Ellen Wilkinson Building, Oxford Road Manchester M13 9PL, UK
e-mail: david.swanson@manchester.ac.uk

B. Doig et al. (eds.), *Interdisciplinary Mathematics Education*,
ICME-13 Monographs, https://doi.org/10.1007/978-3-030-11066-6_10

tinct subjects operating around a shared problem or project; and project work where individual subjects begin to lose their distinctions to various extents.

The previous survey also stressed how little research had been done in the field of inter-disciplinary mathematics, and that what did exist was dominated by small-scale case studies, often with a lack of conceptual clarity, whether terminologically or theoretically. The chapters in this section, therefore, bring very welcome new light and further clarity in a range of areas. This is not only in relation to previously more typical small-scale examples of interdisciplinarity found in practice, but also in relation to the developing existence of longer-term practices, and those that go beyond local to regional, or even (national) system-wide activity or innovation. The scale and form of the interdisciplinary activities involved include: an established national qualification, a long-running regional project at the edges of schooling, a multi-school project integrated into mainstream curricula of different subjects, and a single lesson sequence in one classroom.

In this introductory section I will first outline some of the key features of the forthcoming chapters. Then, drawing on some of the emerging themes, I will focus on an important distinction between mathematics as a tool and mathematics as a (conscious) generalisation. This aims to contribute an essential understanding that can begin to help overcome some of the limitations that can arise when mathematics joins with other subjects (both limitations for the learning of mathematics, but also for learning in areas of life and schooling outside of mathematics, whether in projects, problems or other subjects).

10.1.1 The Case Studies

In the first chapter of this section, den Braber, Krüger, Mazereeuw and Kuiper discuss a successful interdisciplinary STEM course in upper secondary school in the Netherlands, taking place in 40% of the schools there. The course on *Nature, life, and technology* is built around real world problems, and although separate from other subjects, it is staffed by a team of disciplinary teachers drawn from core subject teachers in mathematics, chemistry, biology, physics, and geography. Mathematics appears to have an exceptional role within the course, given one of only four key characteristics expected to be made visible through the curriculum, is 'the role of mathematics in science'. The authors explore whether the stated objectives of the course in relation to mathematics are reflected in practice and the experience of teachers.

Findings indicate diversity in how mathematics operates within the course. In the curriculum the emphasis is on the use of mathematics rather than the learning of specific mathematical concepts, and seems to prioritise procedural over conceptual or strategic mathematical thinking. How this plays out in practice depends on the role of mathematicians within the team. However, although there are some positive stories, in practice only 50% of the teams have a mathematics teacher within them, despite this being named as a priority from the beginnings of the course, and generally students see mathematics as having a lower level or lesser role. The authors discuss the obstacles to mathematics teacher participation, the challenges that those teachers face when they take part, and ways they might overcome these difficulties.

There is a clear contrast between teachers' stated views on the importance of mathematics to the course, whatever their own subject, and that of students and what is often seen in practice. This may be partly due to mathematics importance being seen in its role as a tool for other subjects rather than its full richness. The authors touch on this question, and the contrasting perspectives of mathematics as a servant or queen of science. I will return to this question in the latter parts of this introduction and suggest a more helpful perspective that is of more use in interdisciplinary work—that of mathematics as a particularly helpful (conscious) generalisation of science.

Our next chapter, by Gorriz and Vilches, explores some delightful interdisciplinary work in secondary schools in Catalonia. The authors work with schools across the region on a variety of cross-curricular projects including the making of films using special effects, the design of packages of fixed volume but with shapes designed by students, and the creation of musical instruments.

What is different in their approach, to many others, is less in the bringing together of teachers in different subjects to design, collectively, the central activity, but more in the fact that teachers in each individual subject address central or related aspects of the task within their own classes, with close attention and alignment to the prescribed curriculum in each area. Common co-operative working guidelines are drawn up and then working guidelines within each subject. Student activities are then assessed through the writing of reports within each subject, and overall. So, for example, in the design of packaging the mathematics class focuses on the variations of complexity in particular shapes, and the calculation of volume, sometimes with computer assistance. The authors show how the activities lead to the development of interesting mathematics in the classroom. Meanwhile in science, for example, the nature of the food content is explored; in art, the aesthetics of the packaging is designed; in linguistics, the content of an advertisement is written; and in music, a jingle for the advertisement is created.

The authors argue that such activities are successful for students because of the consistency between subjects, the relation of education to problems close to the heart of students, the emphasis on collaborative work, and because students can approach the tasks at their own level. Teachers also gain much from the activity, although the time required for joint activity is seen to put, potentially, limits on its spreading without sympathetic teachers, or perhaps structural changes. Apart from this burden of additional time for teachers, this model of interdisciplinarity offers much potential due to its accommodation rather than challenge to existing curricula. How much that accommodation puts limits on the genuine integration of subjects, and whether that matters, is a topic for future debate.

In our third chapter, Hobbs, Doig and Plant discuss a regional project based in Victoria, Australia which aims to address the relatively low take-up of STEM qualifications in the area. The project, which engages teachers from 10 schools in a 2-year program, is unusual in its flexibility and range of outcomes. The schools work collaboratively with the local University to develop new knowledge, language, pedagogy, and curriculum, to support the development of the school's own 'STEM vision'. Cycles of professional development including intensive sessions, and support in school are developed on the basis of ongoing assessment and negotiation between the university and schools.

This process has led to a broad range of approaches to STEM and interdisciplinarity which are outlined in the chapter, alongside case studies which provide some rich illustrative detail of the diversity of results. Despite the differences, what unites them is their non-traditional pedagogic approaches, particularly through problem solving but also through strategies such as group work, peer-teaching and open-ended investigations. However, the important emphasis on outcomes such as engagement and creativity is seen to sometimes sideline conceptual development, and the dominance of particular subjects sometimes means that mathematics (and even science) are reduced to being mere tools.

Our fourth chapter, by Kastberg, Long, Lynch-Davis and D'Ambrosio, explores an example of inter-disciplinarity based around an art lesson for 10 and 11 year olds, but one where mathematics inspires and informs the activity, and students begin to develop a new, more positive, relationship with mathematics through the artistic task. The lesson, where groups of students are given the problem of enlarging artwork collectively, is argued to act as a counterpoint to the more common negative experiences of failure and alienation from the subject, partly through its disguising of mathematics in a fun and creative activity, but also through the way the activity allows the re-configuring of relationships, including peer-to-peer, student-to-teacher, and student-to-mathematics.

The space opened up by a focus on collaborative art and creativity encourages students to begin to shift authority and decision making from the teacher to their own collective discussions and activity. As part of this shift students begin to determine both their own working relationships and criteria for success. Students develop important aspects of proportional understanding in, and through, working this way. These are primarily perceptual aspects though, and the authors discuss the complex decisions to be made in such situations, where looking for opportunities to develop mathematics more explicitly may lose the gains that artistic and collaborative work offer.

This chapter therefore helps us to understand better the potentials within mainly mono-disciplinary contexts for teaching that can escape its boundaries and assist in the development of understanding in other disciplines, while still maintaining the positive features of the original discipline. It also raises questions about the limitations of such contexts to go further without losing those positive features. But perhaps, above all, it encourages us to imagine a less alienating school mathematics, where creativity, collaboration, and autonomy feature as much as they can in art.

In the final chapter in this section on case studies, Doig and Jobling provide us with a more general discussion on interdisciplinarity. Their chapter integrates a range of case studies to place the trend for STEM integration in its recent and more long-term historical context. In doing so, they provide a strong argument for the importance of project-based learning as an effective form of interdisciplinarity.

Early examples of interdisciplinarity include the post-war integration in schooling of the mathematics specifically required for manual trades. More modern examples given, such as the long-running Model Solar Vehicle Challenge in Victoria, arguably continue this lineage with an emphasis, for example, on technological design and electrical and mechanical engineering, providing experiences similar to those found within employment in those careers. Such recent examples are argued to be more commonly found on the edge of schooling rather than fully integral and the chapter explores a variety of existing forms to ground its analysis of the potential for making project work, and interdisciplinarity generally, more central to curricula.

The role of mathematics within the integration attempts discussed is not without its problems. In an example looking at mathematisations of acceleration, the expected mathematical model was found at times to complicate rather than assist the development of understanding. One survey of STEM integration projects discussed suggests that such projects are less effective in terms of student attainment gains when mathematics is included alongside other subjects. And finally, it is shown that there may be a tendency for mathematics to only be included in project-based learning in its limited role as a tool, for example, through measurement.

This last chapter therefore highlights some key points that arise in most of our other case studies. We find that engaging with real world problems (in project work, interdisciplinary problems, or art) brings great benefits for the learning of mathematics. It provides substance and meaning, but also engagement and many other things, including even joy, to mathematics. At the same time it also seems to take something away, partly through the potential for losing any conscious focus on mathematics. Through that we may be left with mathematics as a tool, and ultimately lose other benefits, even for the problems at hand, that would come from a richer and more conscious understanding of mathematics, and its own system and connections. One solution, seen within one of our case studies, (by Gorriz and Vilches) may be that alongside bringing subjects together, we may also need a separation, with space to develop mathematics for itself, that can then feed back into more general problem solving. But is this necessarily so? In what remains of this introduction I will focus on these issues.

10.1.2 Mathematics as Tool and Mathematics as (Conscious) Generalisation

> I do not trust teachers of other disciplines to be able to tie the bonds of mathematics with reality which have been cut by the mathematics teacher. (Freudenthal, 1971, p. 420)

In this slightly surly quote, from the irrepressible Han Freudenthal, his main concern is the damage done to mathematics (and perhaps learning in other disciplines) by the subject's separation from the real world contexts and practices that other subjects might provide. One can disagree with aspects of this quote, for example it seems to be blaming individual teachers whereas the forces that have led to the structural separation of the school curriculum into subjects seem beyond any individual. The agreement one can find with Freudenthal though, through studies of interdisciplinary

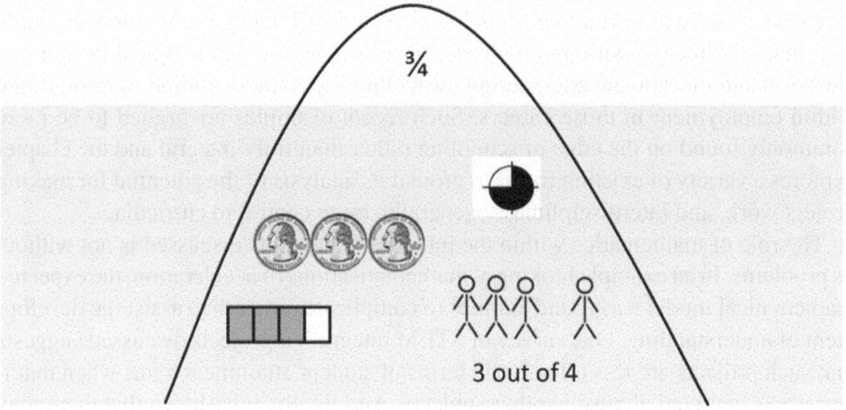

Fig. 10.1 A simplified version of the RME iceberg metaphor

work, including the case studies featured here, is on the essential problem that flows from the artificial separation of the subjects. Mathematics loses the rich concrete material that is an essential element within its abstractions.

In Realistic Mathematics Education (RME) (see Van den Heuvel-Panhuizen & Drijvers, 2014) which builds on Freudenthal's work, there is a popular metaphor of an iceberg, where a mathematical concept such as ¾, resting above the water, requires a range of contextual experiences and examples beneath the surface to support it (see Fig. 10.1).

This image on its own however could fit equally well with an idea of mathematics as a tool. Although ¾ here provides a generalisation across a range of experiences and contexts, it is not necessarily a conscious mathematical generalisation. There is little need for consciousness of what ¾ means as a mathematical thing, tool or object in order to use it. For it to become conscious requires something else to come in. Here, I borrow from Vygotsky's (1987) understanding of scientific concepts, which involves an understanding that abstractions and generalisations do not represent a shedding of concrete experience (typically, abstraction is viewed as removing such concrete details to leave formal mathematical objects), but instead the bringing in of the systemic relationships between concepts, including generalisations of generalisations. This is a complex idea, so let's simplify it (further) by adding something to our original iceberg metaphor.

Above the new line (see Fig. 10.2) lie the systematic connections with other mathematical concepts, such as the relationship between vulgar fractions and decimal fractions, equivalence of fractions, and ratio and proportion in general *et cetera.* This shift in focus is connected to that described as involving a shift from being an activity at one level to being an object of reflection at a higher level (Freudenthal, 1991, p. 96,—and it is therefore worth adding that this general philosophical point is not absent from either Freudenthal or RME, despite being absent from this visual metaphor to help illustrate RME).

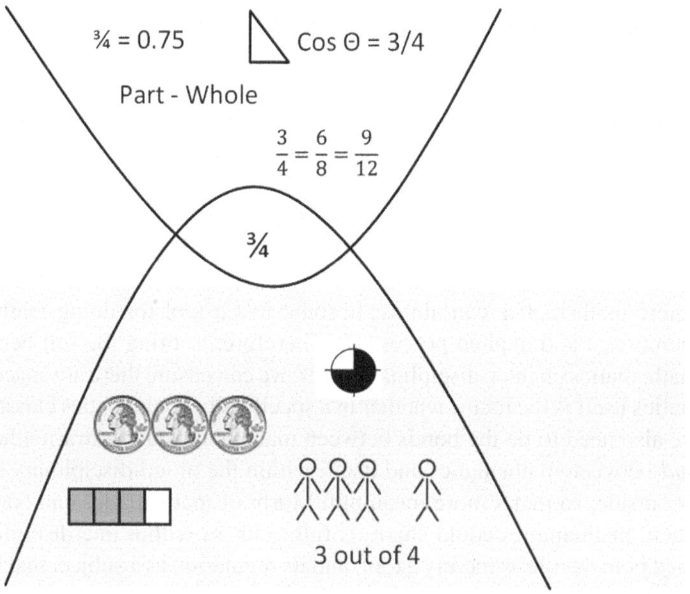

Fig. 10.2 A simplified extension of the RME iceberg metaphor to include mathematical connections

What the introduction of systematic relations between concepts does, in the image, as well as for any individual, is to carve out ¾ as a conscious mental object. This is one that can be picked up and used more readily, and one that becomes a point of conscious connection between the systematic relations of mathematics and any particular example, or experience. (I should add here the caveat that there are of course many ways to challenge this simplistic picture, e.g. can we really distinguish between things above one line and below the other, i.e. can they play similar roles, or, even, are there no objects at all but only processes and relationships? Still, I would suggest it is a useful initial metaphor).

In mathematics education, such generalising, and making of connections, plays a vital role in developing understanding of any particular mathematical activity (e.g. generalised understandings of proportion can mediate understanding of a whole range of topics usually separated in schooling). Mathematics, and systematic mathematical knowledge, can also play a role for the sciences, not just as a tool, but as a particularly useful conscious generalisation of scientific knowledge. It can provide a form of generalisation that unites a variety of physical relationships (e.g. the mathematics of proportion can be seen in Ohm's law, the wave equation, and distance, speed, time relationships) and this generalisation can then mediate and enrich understanding of any of the particular relationships.

This understanding helps us see how mathematics can be of use in interdisciplinary work beyond its role as a tool, but for this to happen this requires some space for specifically mathematical connections to be made (even if in parallel with the introduction of new contexts and experiences). One solution may therefore be that hinted at in the chapter authored by Gorriz et al., where cognate subjects remain distinct even while a joint project is developed. However there is a potential problem with this. To build on the original quote by Freudenthal above, school mathematics, in general has not only cut the bonds between mathematics and reality, but also between mathematics and itself. Curricula are often divided, and divided again, into small bite-size pieces, which perhaps aids memorisation, but certainly undermines understanding. So we find that the aspects of mathematics which are missing from, for example, project work, are also often missing in the mathematics classroom, where mathematics can almost become just a tool for doing mathematics (i.e. in memorised and applied processes). Therefore, to bring the full benefits to bear, of mathematics on inter-disciplinarity, yes, we can ensure there are spaces where mathematics itself is the focus, whether in a specifically mathematics classroom or not. But we also need to tie the bonds between mathematics and extra-mathematical reality, *and* between mathematics and itself, within the mono-disciplinary subject as much as outside, so that a more meaningful form of mathematics can come to dominate. Then, mathematics could share its full richness within interdisciplinary work, and shed both its role as merely a tool, and its reputation as a subject that brings too many negatives along with it.

References

Freudenthal, H. (1971). Geometry between the devil and the deep sea. *Educational Studies in Mathematics, 3*(3), 413–435.

Freudenthal, H. (1991). *Revisiting mathematics education: China lectures*. Dordrecht: Kluwer.

Swanson, D. (2016). Expressions of the commodity form: Alienation and mathematics education. In H. Straehler-Pohl, N. Bohlman, & A. Pais (Eds.), *The disorder of mathematics education*. Springer

Swanson, D. (2018). Vernacular: A montage of mathematical graffiti, quotations and commentary. *The Mathematics Enthusiast, 15*(1), 133–147.

Swanson, D., & Black, L. (2017). Integrating critical theory and practice in mathematics education. In *Proceedings of CERME 10* (pp.1569–1576).

Van den Heuvel-Panhuizen, M., & Drijvers, P. (2014). Realistic mathematics education. In S. Lerman (Ed.), *Encyclopedia of mathematics education* (pp. 521–525). Netherlands: Springer.

Vygotsky, L. S. (1987). Thinking and speech. In *The collected works of L.S.Vygotsky*, Vol. 1. New York: Plenum.

Williams, J., Roth, W.-M., Swanson, D., Doig, B., Groves, S., Omuvwie, M., Borromeo Ferri, R., & Mousoulides, N. (2016). *Interdisciplinary mathematics education: A state of the art (ICME-13 topical surveys)*. Dordrecht, The Netherlands: Springer.

Chapter 11
Mathematics in an Interdisciplinary STEM Course (NLT) in The Netherlands

Nelleke den Braber, Jenneke Krüger, Marco Mazereeuw and Wilmad Kuiper

Abstract Mathematics is one of five disciplines in the successful interdisciplinary STEM course 'nature, life and technology' (NLT) for upper secondary education in the Netherlands. In this study the way mathematics manifests itself in the course is examined at the intended, implemented and attained curriculum. The results show that students often don't recognise mathematics in the course or that they consider the level of mathematics in the course as too low, not interesting. Besides this, mathematics teachers seem to struggle more with their role in interdisciplinary teaching than science teachers. Some feel that there is no need for their mathematical expertise, due to the extent to which mathematics is used. Even though the course has many positive outcomes when it comes to mathematics in the course NLT not only shows diversity in practice but also raises questions we need to address if we want mathematics to be part of interdisciplinary education.

Keywords Mathematics · Interdisciplinarity

11.1 Introduction

An interdisciplinary STEM (Science, Technology, Engineering and Mathematics) course can provide opportunities for teachers to collaborate, show students cohesion between the disciplines and use real-life problems as the starting point of investigation. In the Netherlands, such a course exists with many good practices that utilize

N. den Braber (✉) · M. Mazereeuw
NHL Stenden University of Applied Sciences, Leeuwarden, The Netherlands
e-mail: nelleke.den.braber@nhl.nl

M. Mazereeuw
e-mail: m.mazereeuw@nhl.nl

J. Krüger · W. Kuiper
Freudenthal Institute, Utrecht University, Utrecht, The Netherlands
e-mail: jenneke.kruger@gmail.com

W. Kuiper
e-mail: w.kuiper@uu.nl

B. Doig et al. (eds.), *Interdisciplinary Mathematics Education*,
ICME-13 Monographs, https://doi.org/10.1007/978-3-030-11066-6_11

these opportunities. However, when it comes to mathematics, in the course NLT, research not only shows diversity in practice but it also raises questions that need to addressed if we want mathematics to function within interdisciplinary education. In this paper, we describe the case of NLT to discuss these issues.

11.1.1 Background of NLT

In 2007 the course 'Nature, life and technology' (NLT) was introduced in the curriculum of upper secondary education in the Netherlands for students aged 16–18. The general aims of this interdisciplinary STEM course are to increase the attractiveness of science education for students and to let students experience the importance of interdisciplinary coherence in the development of science and technology (Stuurgroep NLT, 2007). The course is intended to supplement the existing disciplines in the Dutch curriculum: physics, chemistry, mathematics, biology and physical geography. It aims to offer both a broader and more in-depth educational programme for science and mathematics and is not meant as a replacement of other courses.

Schools are not obliged to offer NLT. About 230 schools do at the time of research (2015), which is approximately 40% of the schools in the country. For students, it is an elective course, chosen additionally to other courses and assessed by a school-based exam. More than 6000 students finish the course each year.

NLT is designed to offer more freedom for teachers and students. It allows them to choose what they find interesting and to use the potential of their local region with relevant contexts for learning, for instance, in visiting research institutes or local industry. Within the boundaries of the examination programme, teachers can select teaching materials from a wide range of compact booklets. These materials are called *modules*. Each module introduces a contemporary science problem that can only be solved by involving different (disciplinary) perspectives resulting in a context-orientated and interdisciplinary course. As a consequence, it is desired, and necessary, that the course is taught by a team of teachers, called an NLT team. Preferably one of each of the intended disciplines is represented in the team.

As described by Michels and Eijkelhof (2014) all modules were specifically developed for the course through co-operation of secondary school teachers and experts from universities, colleges, research institutes and, or, industry. The rich variety of modules is meant to contribute to student awareness of the possibilities for further education in a scientific or technological area.

During the first ten years (2006–2015) the development and implementation of NLT was funded by the Dutch department of education. In 2016 schools formed an association of NLT schools to promote further development and to maintain the quality of the course.

11.1.2 *Focus of the Study*

From the start of NLT, it was made clear that

> Attracting mathematics teachers to participate in such a team has priority because mathematics plays the rôle of language and/or tool in many sciences. (Stuurgroep NLT, 2007, p. 8)

In the examination programme of NLT, the nature of the course is made explicit by formulating four characteristics that should be visible throughout the curriculum (Krüger & Eijkelhof, 2010). The nature of NLT characterizes itself by 'interdisciplinarity', 'the relationship between science and technology', 'the orientation on higher education and occupations', and 'the rôle of mathematics in science'.

By denoting the rôle of mathematics as one of the main characteristics of the course a unique situation occurs. Although NLT aims to make clear to students that many different disciplines are needed to solve a problem without suggesting that any one discipline is better than another, the rôle of mathematics seems to need more emphasis. Research activities in the last few years, such as interviews with teachers, students, and developers, show that mathematics often has an exceptional position. Examples can be found in the development and use of the teaching materials and the participation of mathematics teachers in an NLT team. In interviews with NLT teachers, mathematics in NLT is often described as a 'different' story. The study presented in this paper focuses on that story.

11.1.3 *Research Question*

In this exploratory study, we investigate whether the objectives of the course, as described in the formal vision document (Stuurgroep NLT, 2007) and examination programme (Krüger & Eijkelhof, 2010) are met when we concentrate on the mathematics in the course and the participating of mathematics teachers. It aims to depict how mathematics functions in NLT by means of examining the current situation at different representations of the curriculum.

The underlying question is

> How are the objectives related to mathematics reflected in the curriculum of the course NLT?

11.2 Conceptual Framework

Rationales for interdisciplinary STEM courses are often based on the fact that the problems we face in today's world call for perspectives and knowledge from many different areas. The possibilities for mathematics in such a course is described by Williams et al. (2016, p. 13) as interdisciplinary mathematics education offers mathematics to the wider world in the form of added value (e.g. in problem solving), but on the other hand also offers to mathematics the added value of the *wider world*.

At the same time they mention that some interdisciplinary literature points out that it seems mathematics is gaining the least from integration.

Evidence for learning gains from interdisciplinary work mainly concerns students' engagement, motivation and problem solving skills (Czerniak, 2007). Even though some results show gains in achievement scores (Czerniak & Johnson, 2014) the outcomes *that will likely be affected by interdisciplinary working will be non-traditional, and non-standard* (Williams et al., 2016, p. 17). For NLT, compared to the examination programmes of the mathematics school curriculum, non-traditional and non-standard outcomes are formulated. For instance, there are no learning outcomes that describe the learning of specific mathematical concepts but they do describe the use of concepts relevant to the interdisciplinary context (Krüger & Eijkelhof, 2010). The same can be said for concepts from physics, chemistry, biology and geography. Also skills and methods required in interdisciplinary working are formulated in the examination programme. This is consistent with the fact that NLT is a context-orientated course in which authentic professional practices, or problems, are the starting point of a module. Something students seem to find meaningful (Dierdorp, Bakker, Maanen, & Eijkelhof, 2014) but is very different from the way mathematics is taught in the Netherlands in general.

When mathematics teachers start teaching NLT they are confronted with differences in the teaching practice between NLT and their familiar mathematics course. Aims, course structure, learning activities, pedagogy, and content can all be different from what they are used to working with. Moreover, knowledge of other school courses and pedagogy is required, such as practical work.

There are various factors that have an impact on attempts to implement interdisciplinary curricula (Czerniak & Johnson, 2014; Venville, Rennie, & Wallace, 2012). Some factors, such as curriculum and testing constraints, and lack of suitable materials, have little impact on NLT, or are not specific to the manifestation of mathematics in NLT. However, teacher content knowledge, and teacher beliefs and attitudes, seem to need consideration when implementing NLT as NLT teams consist of teachers with very different backgrounds.

This is supported by the study of Rıordain, Johnston and Walshe (2015). Teacher perspective, teacher knowledge of the 'other subject', technological pedagogical content knowledge (TPACK), and teacher collaboration and support, were found to be key aspects that have an impact on the integration of mathematics and science teaching and learning. Strategies for collaboration and team work as well as a deeper understanding of content across STEM, that can boost such integrations, are also mentioned as something that should be included in teacher education programmes (Berlin & White, 2012).

In accordance with the teachers' perspectives, students perspectives on NLT and the position of mathematics in NLT is relevant, as NLT aims to show students how mathematics is used in science and technology (Krüger & Eijkelhof, 2010). One of

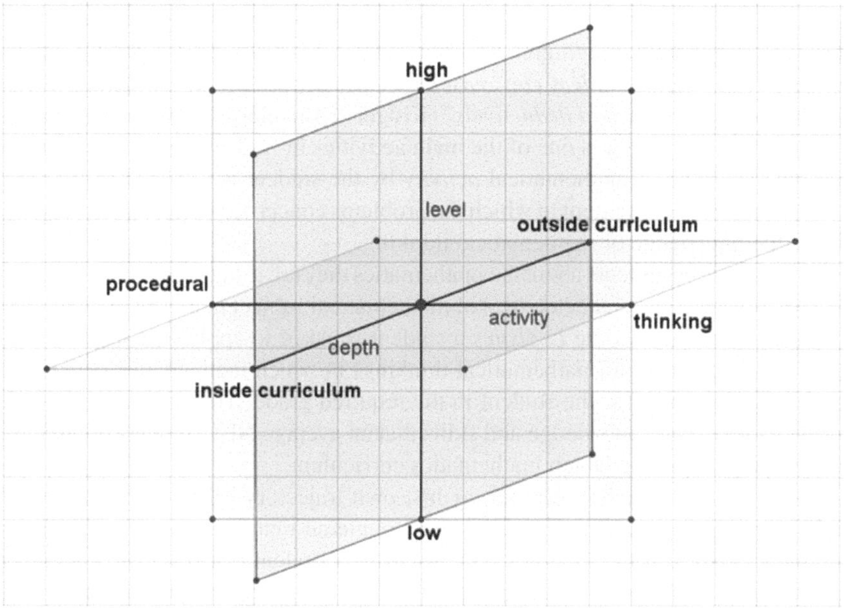

Fig. 11.1 Possibilities of mathematics encountered by students in NLT

the factors that influence students' perspectives is the characteristics of the teaching materials. We need an analytical framework to see how mathematics manifests itself in the teaching materials and to position the perspectives of the students. The framework we have constructed is shown in Fig. 11.1 where three axes represent three aspects of the possibilities of mathematics a student can encounter in the teaching materials when mathematics is required to solve a problem.

The view that NLT should contain both the use of known and unknown mathematics (Stuurgroep NLT, 2007) is reflected in the horizontal axis *(measure of) depth*. When mathematical fields are required that are not part of the school mathematical curriculum, for instance graph theory, this is characterized as *outside the curriculum*. Related subject matter is presumably unknown for students. This also applies to additional content on a known topic such as differential equations as an extension of calculus. The encountered mathematics is positioned on the axis by the frequency in which content is presented in the mathematics curriculum or the connectedness to the existing curriculum (cTWO, 2012).

Mathematics encountered in a module can also involve that which is taught in the mathematics curriculum at a higher or lower grade. The vertical axis *level*, is used to position the offered mathematics compared to the current mathematics curriculum of the student (cTWO, 2012).

Since 2012 new examination programmes for the sciences and mathematics courses have been implemented in the Netherlands. The development of these programmes resulted in a renewed focus on mathematical thinking in the Dutch curriculum (cTWO, 2012; Drijvers, 2015), specifically on problem solving and modelling. The NLT examination programme does not address this distinction explicitly, except for the phrase '*that a student can reason consistently with mathematical and scientific data inductively and deductively*' (Krüger & Eijkelhof, 2010, p. 38), and the statement that modelling is one of the main activities in NLT along with design and research. The type of mathematical *activity* by the student is reflected in the third axis that represents the extent to which the problems concerned require mathematical procedural knowledge or mathematical thinking.

Perspectives of students about the mathematics they are using can be positioned on the axes as well as the characteristics of mathematical subject matter in the modules. For instance, in the module *Logistics* a student is asked to apply new mathematics to a modelling situation (mathematical thinking) in which mathematics is required at a level appropriate for the student in the required grade. This statement is made based on the average knowledge and skills that an average NLT student should have at a certain grade in the Dutch mathematics curriculum.

Obviously, each student has her, or his, own trajectory of development which could mean that personal perceptions of encountered mathematics can differ from the characteristics of the modules. For instance, a student can perceive something as unknown mathematics when in fact it has been offered to the student in the mathematics curriculum.

11.2.1 Method

The different representations of the curricula are used as a framework to explore the current NLT curriculum. The distinction between intended, implemented, and attained curriculum (Goodlad, 1979; van den Akker, 2003), has shown to be especially useful in the analysis of the processes and the outcomes of curriculum innovations (SLO, 2009). At the different representations, data collection has focused on teachers, students, content, and teaching materials.

11.2.1.1 Data Collection

To answer the question on how the mathematics objectives are reflected in NLT, data was collected on the intended, implemented, and attained curriculum. Data on the intended curriculum has been derived from an analysis of (non) official documents from the period of early development of the course, teacher examination guides, as well as from interviews with developers.

Data on both the implemented and attained curriculum was collected through interviews, surveys, and a content analysis of modules. All teaching materials are available for research as well as user data through an on-line user database. Interviews with NLT teachers and students (27 teachers, 50 students) were held while visiting nine NLT schools in 2013. During these school visits many NLT related topics were covered, and the relationship between mathematics and NLT was one of them. There were three interviews with teacher educators in 2014. In 2014 and 2015, students were asked to complete a survey. These students were all in their final school year to increase the probability that they had a good overview of the whole course. All in all, 914 student surveys were completed. Finally, in 2015 two teacher surveys, one among NLT teachers and one specifically for mathematics teachers provided data from 254 teachers, 70 of whom taught mathematics.

Additionally, data for the attained curriculum consists of NLT registration forms of 202 schools and the grades of all NLT students from the national school administration organization (DUO). The registration forms contain specific information about the number of teachers that participate in an NLT team.

11.2.1.2 Instruments

The questions in the semi-structured interviews and the teacher surveys focused on the perceived relationship between mathematics and NLT, reasons for participating in an NLT team, and the importance of mathematics in NLT. Questions corresponding to these categories were for example '*how do you see the rôle of mathematics in NLT?*' or '*why should a mathematics teacher (not) participate in NLT?*' and '*do you think it is important that mathematics is a part of NLT and state your reasons?*'. Students were asked to describe NLT, which disciplines they perceived as a part of NLT, and what their opinion of the course was. They were also asked about the mathematics in NLT. The questions were related to the possibilities in Fig. 11.1 as they focused on the occurrence of unknown mathematics and the difficulty of the mathematics. They were also asked if they could give examples of used mathematics, the visibility of the mathematics, and if they were taught by a mathematics teacher during the NLT course. The two student surveys were identical except for one aspect. Students that did not perceive mathematics as a discipline in NLT were only asked one more question in the first survey, namely why they thought mathematics was not a discipline in NLT. In the second survey students were asked to complete all questions to see whether they were consisted with their perception of mathematics not being a discipline in NLT.

11.3 Data Analysis

The data on students and teacher materials was categorized using the different ways of the occurrence of mathematics in NLT, as shown in Fig. 11.1, and the knowledge of both the mathematics and NLT curriculum and practice of two of the authors. For instance, when a student claims that he only encountered basic calculations this is categorized as low *level* mathematics or when a student describes that he needed to know how to solve linear equations this is marked as a procedural *activity*. The responses of students from both surveys were combined and counted on specific questions related to our framework; for instance, if they encountered unknown mathematics.

The teacher surveys and interviews were analysed by examining teachers' perspectives on participation in NLT teams as well as their perspectives on NLT, the position of mathematics in it, working with the materials and collaboration with teachers from other disciplines.

11.4 Results

Below, we present a few results of our analysis. Starting at the objectives of the course we gathered data on the different representations of the curriculum by looking at course and teaching materials and the view-point of teachers and students. Results are ordered accordingly.

11.4.1 NLT Curriculum

In the examination programme (2010) descriptions referring to mathematics were added to each of the examination domains e.g. *the student can apply relevant science and mathematical concepts to* [...]. However, additional documents such as the teacher examination guide show that the rôle of mathematics is limited to a list of techniques that students can encounter in the modules, many of these techniques have already been taught in lower secondary education. That is, these techniques are mentioned in objectives and explained in course materials of lower secondary education.

NLT, and thus the mathematics within it, is intended to supplement the mathematics curricula. However, NLT students have different backgrounds in mathematics, depending on which mathematics programme they have attended when entering the course. Some have experienced no statistics or advanced calculus. Others have attended an advanced mathematics course, called math D, which covers topics like dynamic modelling and complex numbers, and is also an elective school based examination course. As intended by the developers of both NLT and math D, modules were developed that could be used in both courses. In some schools these modules are indeed part of the NLT curriculum, although sometimes presented as isolated math D modules and not as NLT. This could be a way to ensure that more mathematics is offered to students if the school has decided not to offer math D.

11.4.2 Teaching Materials

Given that the modules are context-based, the required disciplinary knowledge varies between modules. Some require more mathematical skills than others. Every school has its own selection of modules and therefore school curriculum. The only condition is that the examination programme of each student is met by the selected modules. To help schools to arrange their curriculum a group of experts has described all used disciplinary concepts in all modules (voorbeelden en gereedschappen, n.d.). This data shows that there are 21 (out of 72) modules that require minimal mathematical knowledge or only mathematics from lower secondary education (4 out of 72 modules). Comparing the educational materials of NLT with the examination programme has shown that it is possible to build a suitable curriculum with a selection of these modules.

There are 18 modules that use concepts which are not part of the compulsory school mathematics curriculum; e.g. linear programming, graph theory, logic, or differential equations. Eleven modules use statistical, or probability theory, which is not in the basic curriculum of the majority of the students.

Procedural activities involving solving equations, working with formulae and calculus are more common than activities involving mathematical thinking.

Modules that have a strong mathematics component are not represented in the five most popular modules (in 2014), except for one, dynamic modelling. This module deals with a topic that is not part of the mathematics school curriculum and was developed to be used in the advanced mathematics course mathD as well. In Fig. 11.1 such a course is positioned on the end of the axis *Depth* and on the procedural side of the axis *activity* near the centre. Depending on the mathematical courses students take, the level of mathematics varies from average level to high level. This means that encountering new mathematics does not necessary mean that a high level of mathematics is required.

11.4.3 Teachers in NLT

Mathematics teachers are allowed to teach NLT. However, not all mathematics teachers are familiar with NLT, or they are not aware that they are allowed to participate in a team of NLT teachers. Interviews and informal conversations with teacher trainers show that this is also true for teacher trainers. There doesn't seem to be a general focus on NLT under mathematics teachers. This is supported by the fact that even though 200 (20%) of NLT teachers responded to the first teacher survey only 18 of them were mathematics teachers. And only a small number of mathematics teachers completed the second survey sent to more than 4000 recipients of a popular educational newsletter, most of them mathematics teachers. The completed surveys combined with the interviews do, however, provide insight into the participation of mathematics teachers.

Table 11.1 Number of teachers in NLT of the mono disciplines from 202 schools

Course	Percentage of schools with teacher(s) of mono discipline in NLT-team (%)	Total number of teachers in NLT teams
Geography	40	93
Biology	95	291
Physics	98	288
Chemistry	98	255
Mathematics	50	121
Other (e.g. computer science, physical education)	9	19

The reasons mathematics teachers have to participate vary. Teachers who wish to participate in NLT indicate that NLT shows attractive applications of abstract mathematics, and ways to make mathematics and its use, visible for students. As one mathematics teacher says:

> The question 'where do you use mathematics' is finally answered.

Also, working together with other NLT teachers is mentioned as a positive aspect of NLT. Some of the mathematics teachers recommend participation in NLT, for the course itself, but also for their own professional development; '*it broadens your own view*'. Others feel less strongly about the necessity to participate. On the contrary, the opinion that science teachers are sufficiently able to help with the mathematics in the course is mentioned as a reason for not wanting to participate. As described by a teacher and shared by others:

> Mathematics is actually quite scarce in the course and when you encounter it, is superficial. Often a teacher of a science subject can also explain it to the students.

Other reasons for not participating are that teachers feel insecure about their science knowledge, that they want to concentrate on mathematics lessons, or that they are not allowed by their school administration, because of a shortage of mathematics teachers.

On average, five teachers participate in an NLT team. But in only 50% of the teams a mathematics teacher is present, as shown in Table 11.1. In contrast, in 98% of the NLT teams at least one physics teacher is present. Even though teacher formation in schools may vary every year, great shifts in numbers are not to be expected.

The low participation of geography teachers in NLT can be explained by the availability of geography teachers with knowledge of physical geography, which is only a small part of the geography curriculum.

Some schools ask teachers to participate in NLT for specific modules because of their expertise in certain fields like computer science, engineering, or physical education. In Table 11.1 these teachers can be found under *Other.*

The question *do you think it is important that mathematics is a part of NLT and why* was answered positively in the surveys by 56 of 64 mathematics teachers. Not all explicated their answer and reasons for finding mathematics important, and even though the answers vary, they seem to coincide with the reasons to participate.

Some answers to illustrate the diversity:

> Mathematics is necessary to fully comprehend scientific models. Mathematics is used everywhere but that is not visible in the mono-disciplines. There is no point because it is such a small part. Mathematics should be stressed more as a separate discipline.

11.4.4 Students

Analysis of the student surveys show that students have very different opinions about mathematics in NLT. Although most students recognize mathematics as a discipline within NLT, 22% don't mention mathematics when asked about the disciplines that play a rôle in NLT. In contrast, 98% mention physics as playing a rôle. When asked why mathematics is not mentioned, students say that mathematics plays a smaller part than the other disciplines. They also say that only low-level mathematics is required, or that it is not similar to what they do in mathematics class. Several students mention that it coincides more with what they do in physics lessons.

> Question: You didn't choose mathematics as one of the disciplines in NLT. Can you describe why not?
>
> Student 1: Because there is little maths in it. If it is there, it's basic calculations that you apply in physics or geography.
>
> Student 2: You never get maths questions. I didn't even have to differentiate once in the NLT course.

From the 131 students in the second survey who say that mathematics is not a discipline in NLT, 101 say that there was no mathematics teacher part of the NLT team, or that they don't know if that was the case. Also, 22 (out of the 131) say they have encountered mathematics they hadn't learned before.

At the same time, there are students who are very positive about mathematics in NLT. Two examples are:

> Due to NLT I now know that mathematics is a lot more than what is done in mathematics class and now I think about studying mathematics.
>
> Within NLT chemistry, physics and mathematics come together and now I see the use of the three mono disciplines.

A majority (85%), of the students who say mathematics is part of NLT, also indicate that mathematics is important in NLT and that NLT gives them insight in real-life applications of mathematics (73%). About 50% of the students indicate that they have learned mathematical skills and topics that they hadn't learned before in mathematics courses.

11.5 Summary

This exploratory study tried to answer the question how the objectives of the course related to mathematics in NLT are reflected in the curriculum of NLT. Taking the objectives of the course we gathered data on the different representations of the curriculum by looking at course and teaching materials, and the view-point of teachers and students. Data was analysed by looking at how the students encounter mathematics in NLT, and the perceptions and participation of mathematics teachers related to the course objectives.

In the examination programme of NLT, the 'rôle of mathematics in science' is formulated as one of the four characteristics that should be visible throughout the curriculum. However, how this should manifest itself in the course is not clear, and perspectives on mathematics in the course vary both from teachers and students.

The intention of NLT is to offer both a broader, and more in-depth, educational programme than the mono disciplines in school. On the one hand, data shows that NLT calls for a lot of low level procedural knowledge, and that only half of the students encounter mathematics they haven't seen before. Most modules in which new mathematical concepts and techniques are needed do not seem popular among students and teachers, or these are not presented as NLT modules, but as advanced mathematics modules. This may have to do with the fact that NLT students have different mathematical backgrounds. However, that is also the case when it comes to knowledge of physics, biology, or geography.

At the same time, there are students and teachers who feel that NLT is a way to show how mathematics is used, which could be related to the *added value of the wider world,* as mentioned in the conceptual framework.

One of the objectives of the course is to strengthen the cohesion of science disciplines. Although many students see mathematics as an important part of NLT, this is not recognised by all students. Some students feel the mathematics in NLT is not similar to what they encountered in mathematics courses.

At the start of NLT, the participation of mathematics teachers was named as a priority. In practice, we see that, after 10 years, in 50% of the NLT schools no mathematics teacher is present in the NLT teams.

The results call for strengthening the aims of NLT by working towards a well-defined rôle for mathematics in NLT. Further study will address this issue.

11.6 Discussion

The case of NLT provided us with an example of an interdisciplinary STEM course in which teachers from different disciplines collaborate to show students the cohesion between the disciplines, and in which they use real-life problems as the starting point of investigation. The development and implementation structure, government support, and the fact that it is not meant as a replacement of other mono-disciplinary courses, lowers the number of obstacles for integration, as mentioned in literature (Czerniak & Johnson, 2014; Venville, Rennie, & Wallace, 2012). However, when it comes to mathematics in the course, NLT not only shows diversity in practice but the way in which mathematics is manifest also raises questions that need to be addressed if we want mathematics to be a relevant part of interdisciplinary education.

Firstly, the question of whether mathematics teachers should be stimulated to join an NLT team considering that mathematics teachers seem to struggle with their rôle in the course when compared to science teachers. In light of the developments in the field of mathematics, and the general aims of NLT, we argue that the assumed quintessential contribution of mathematics teachers to an NLT team needs to be elaborated.

In 2013, the National Research Council of the United States (NRC, 2013) published a comprehensive analysis of the field of mathematics in which they describe that:

> Mathematical sciences work is becoming an increasingly integral and essential component of a growing array of areas of investigation in biology, medicine, social sciences, business, advanced design, climate, finance, advanced materials, and many more. This work involves the integration of mathematics, statistics, and computation in the broadest sense and the interplay of these areas with areas of potential application.

For the Dutch situation, a similar report was published (PWN, 2014).

NLT teachers have the opportunity to let their students experience *how the mathematical science topics they are teaching are used and the careers that make use of them* (NRC, 2013 p. 11). Here the rôle of the mathematics teachers may come to the surface. By participating in an NLT team, teachers do not only give the example of mathematicians working together with other disciplines to solve problems, they are also the experts on the mathematics curriculum who are able to design learning and teaching processes that take prior knowledge and skills of students into account. Participation provides them with the opportunity to raise the level of mathematics used in the modules, reflect on the functionality of mathematics, and enable the transfer between mathematics and other science disciplines in the NLT curriculum.

Secondly, as described in the conceptual framework, when mathematics teachers begin to start teaching NLT they may be confronted with a teaching practice that, for them, has unfamiliar goals, pedagogy, course structure, and content. The mathematical exercises from mathematics text books are not often seen in NLT. This confrontation requires a change in thinking about the position of mathematics, collaboration with teachers from other disciplines and the use of materials with various natures.

We tend to refer to this ability to change as *agility*. Although all NLT teachers have to be agile in some respect, NLT may ask for more agility of mathematics teachers. When this is the case, support of mathematics teachers in gaining such agility is required.

Visser (2012) described a professional development programme for teachers in a school that can boost team teaching and the professional growth of the teachers in an NLT team. Furthermore, currently some steps are made to describe the competencies of a NLT teacher (Competentieprofiel docent NLT, 2013). However, the best way to help future teachers and specifically mathematics teachers to participate in NLT has not yet been the subject of research. This is not only the case for NLT. Czerniak and Johnson (2014) state that there are still few models about effective ways of preparing teachers to deliver integrated instruction.

Several useful models can be found in studies on the knowledge, beliefs, and attitudes of mathematics teachers, a common division used by Ernest (1989). The focus of these models is on the mathematics teacher in a mathematics classroom. However, in NLT a mathematics teacher with all his, or her, knowledge, beliefs and attitudes is asked to teach in an interdisciplinary course in addition to mathematics courses. Because of this, we wonder what part of the known models can be used to describe the interdisciplinary knowledge, beliefs and attitude required.

Ernest declared that knowledge of the other school courses should be part of the knowledge of a mathematics teacher to provide an opportunity to use other courses to facilitate the learning of mathematics and to show the use of mathematics. In the model describing mathematical knowledge for teaching (Ball, Thames, & Phelps, 2008) it is not clear where this knowledge is positioned. The new component to the existing model *horizon knowledge* described by Jakobsen, Thames and Ribeiro (2013) may well be considered to also include interdisciplinary knowledge but is not yet well-defined. We consider interdisciplinary knowledge as more than knowing where and when mathematics is used in other disciplines, but also how disciplines relate to each other (Repko, Szostakm, & Buchberger, 2014). An NLT teacher should provide an answer to questions like *why is it important that mathematics is a part of NLT, in what way should mathematics be used in NLT* or *what connection to the mathematics taught in mathematics class can be highlighted with this topic?*

This requires not only interdisciplinary knowledge but also relates to personal beliefs about the rôle of mathematics in real-life and in relation to other disciplines. In studies on beliefs, teacher statements related to interdisciplinary teaching are scarce. We argue that the application related view on mathematics teaching (Grigutsch, Raatz, & Törner, 1998; Feldbrich, Muller & Blömeke, 2008), or the utilitarianism in teaching (Tang & Hsieh, 2014), do not cover the complete spectrum. We can use the distinction between beliefs on the nature of mathematics, mathematics teaching and students learning (Cross, 2009) to describe the beliefs concerning mathematics within NLT. For instance, a mathematics teacher in NLT needs to reflect on the importance of mathematics in NLT with respect to the nature of NLT (*nature*), how a mathematics teacher can contribute to NLT, and collaborate with other teachers (*mathematics teaching*), and how students should perceive the rôle of mathematics in NLT (*students learning*).

Attitude towards interdisciplinary teaching also influences participation. In this study teachers gave 'broad interests' or 'preferred focus on mathematics' as reasons to participate in NLT.

Even though we discuss knowledge, beliefs, and attitude, in relation to our case study, we believe that our findings may show similarity to other interdisciplinary STEM courses.

Consequently, further research should focus on how to equip future mathematics teachers for teaching NLT and to meaningfully collaborate in NLT teams. In the end, mathematics teachers in NLT should be able to help students experience the functionality of mathematical sciences in real-life. An extended analysis framework using the concept of functionality in relation to interdisciplinary knowledge, beliefs, and attitude, will be elaborated on in our further research.

When looking back at the results of this study and the discussion above there is one aspect that we have not addressed properly that emerged when reflecting on the outcomes of our study. This is the value of mathematics. Ng and Stillman (2007) describe that value of mathematics as well as mathematical confidence, and the interconnectedness of mathematics are three affective domains directly associated with interdisciplinary learning involving mathematics. They measure the value of mathematics by looking at the perceived usefulness of mathematics by students, i.e. current relevance, usefulness for further education or society. Williams (2012) distinguishes between practical use, purchasing power and enjoyment of mathematics as different themes associated with value. Other themes are social empowerment e.g. learning ways of thinking or having knowledge of our culture (Ernest, 2010).

The way mathematics teachers value their subject is probably closely related to the values mentioned above. The metaphors of mathematics as a servant and queen of other disciplines come to mind. However, the dichotomy of mathematics as queen, or as servant, not only illustrates the problem, the use of this metaphor is in itself a cause of the problem. When mathematics teachers see their own discipline as the queen it is hard to accept that others are perceived to treat it as a servant. Evidently it is not helpful to position disciplines hierarchically. It is more useful to know and appreciate what each discipline contributes to solving real-life problems, and how to get the most out of the characteristic ways of thinking, used methods, and culture of a discipline. For mathematics this could include acknowledging modelling and problem solving strategies, or logical reasoning, next to procedural fluency. Repko, Szostak, and Buchberger (2014) describe this as part of *interdisciplinary perspective taking* where *taking on other perspectives often involves temporarily setting aside your own beliefs, opinions, and attitudes*. (p. 125)

It would be useful for future studies to address interdisciplinary perspective taking specifically how to include this as a part of teacher education. It would be recommended to include aspects of interdisciplinary teaching in national standards for teacher education in countries, like the Netherlands, where such standards exist.

The goal is not marginalising one discipline, or promoting another, but to value each discipline and use the possibilities each discipline brings to the table in a real-life situation.

References

Ball, D. L., Thames, M. H., & Phelps, G. (2008). Content knowledge for teaching what makes it special? *Journal of Teacher Education, 59*(5), 389–407.

Berlin, D. F., & White, A. L. (2012). A longitudinal look at attitudes and perceptions related to the integration of mathematics, science, and technology education. *School Science and Mathematics, 112*(1), 20–30.
Competentieprofiel docent NLT. (2013). Retrieved May 23, from http://www.ecent.nl/artikel/2682/Competentieprofiel+docent+NLT/view.do.
Cross, D. I. (2009). Alignment, cohesion, and change: Examining mathematics teachers' belief structures and their influence on instructional practices. *Journal of Mathematics Teacher Education, 12*(5), 325–346.
cTWO. (2012). *Denken & doen; wiskunde op havo en vwo per 2015*. Utrecht: Universiteit Utrecht.
Czerniak, C. M. (2007). Interdisciplinary science teaching. In *Handbook of research on science education*, pp. 537–559.
Czerniak, C. M. & Johnson, C. C. (2014). Interdisciplinary science teaching. In *Handbook of research on science education* (pp. 395–411). New York: Routledge.
Dierdorp, A., Bakker, A., van Maanen, J. A., & Eijkelhof, H. M. (2014). Meaningful statistics in professional practices as a bridge between mathematics and science: An evaluation of a design research project. *International Journal of STEM Education, 1*(1), 1–15.
Drijvers, P. (2015). *Denken over wiskunde, onderwijs en ICT*. Utrecht: Universiteit Utrecht.
Ernest, P. (1989). The knowledge, beliefs and attitudes of the mathematics teacher: A model. *Journal of Education for Teaching, 15*(1), 13–33.
Ernest, P. (2010). Why teach mathematics? *Professional Educator, 9*(2), 45–47.
Feldbrich, A., Muller, C., & Blömeke, S. (2008). Epistemological beliefs concerning the nature of mathematics among teachers educators and teacher education students in mathematics. *ZDM Mathematics Education, 40*, 763–776.
Goodlad, J. I. (1979). *Curriculum inquiry. The study of curriculum practice*. New York: McGraw-Hill.
Grigutsch, S., Raatz, U., & Törner, G. (1998). Einstellungen gegenuber Mathematik bei Mathematiklehrern. *Journal für Mathematikdidaktik, 19*, 3–45.
Jakobsen, A., Thames, M. H., & Ribeiro, C. M. (2013). Delineating issues related to horizon content knowledge for mathematics teaching. In *Proceedings of the Eight Congress of the European Society for Research in Mathematics Education.*
Krüger, J., & Eijkelhof, H. (2010). *Advies beproefd examenprogramma NLT: eindrapportage Stuurgroep NLT*. Enschede: Stuurgroep NLT.
Michels, B., & Eijkelhof, H. (2014). Professional development of teachers by developing, testing and teaching curriculum materials of an interdisciplinary STEM-course (NLT) for senior high school. In *Conference Proceedings Educating the Educator*, pp. 288–233.
Ng, K. E. D., & Stillman, G. (2007). *Interdisciplinary learning: Development of mathematical confidence, value, and the interconnectedness of mathematics*. Paper presented at the 30th Annual Conference of the Mathematics Education Research Group of Australasia Incorporated (MERGA 2007) on "Mathematics: Essential research, essential practice", Tasmania, Australia, 2–6 July 2007.
NRC. (2013). *The mathematical sciences in 2025*. Washington: National Research Council.
PWN. (2014). *Formulas for insight and innovation. Vision document 2025*. Amsterdam: PWN.
Repko, A. F., Szostak, R., & Buchberger, M. (2014). *Introduction to interdisciplinary studies*. London: Sage.
Ríordáin, M. N., Johnston, J., & Walshe, G. (2015). Making mathematics and science integration happen: Key aspects of practice. *International Journal of Mathematical Education in Science and Technology, 47*(2), 233–255.
SLO. (2009). *Curriculum in development*. Netherlands Institute for Curriculum Development (SLO). Enschede: SLO.
Stuurgroep NLT. (2007). *Contouren van een nieuw bètavak; visie op een interdisciplinair vak: natuur, leven en technologie*. Enschede: SLO.
Tang, S., & Hsieh, F. (2014). The cultural notion of teacher education: Future lower secondary teachers' beliefs on the nature of mathematics, the learning of mathematics and mathematics

achievement. In *International perspectives on teacher knowledge, beliefs and opportunities to learn*. Dordrecht: Springer.

van den Akker, J. (2003). Curriculum perspectives: An introduction. In J. van den Akker, W. Kuiper, & U. Hameyer (Eds.), *Curriculum landscapes and trends* (pp. 1–10). Dordrecht: Kluwer Academic Publishers.

Venville, G., Rennie, L. J., & Wallace, J. (2012). Curriculum integration: Challenging the assumption of school science as powerful knowledge. In *Second international handbook of science education* (pp. 737–749). Netherlands: Springer.

Visser, T. C. (2012). *Professional development as a strategy for curriculum implementation in multidisciplinary science education*. Enschede, Institute for Teacher Education, Science Communication & School Practices (ELAN), University of Twente.

Voorbeelden en gereedschappen. (n.d.) Retrieved May 23, 2017, from http://betavak-nlt.nl/nl/p/lesmateriaal/leerlijnen/voorbeelden-en-gereedschappen/.

Williams, J. (2012). Use and exchange value in mathematics education: Contemporary CHAT meets Bourdieu's sociology. *Educational Studies in Mathematics, 80,* 57–72.

Williams, J., Roth, W.-M., Swanson, D., Doig, B., Groves, S., Omuvwie, M., et al. (2016). *Interdisciplinary mathematics education: State of the art*. Cham: Springer. https://doi.org/10.1007/978-3-319-42267-1.

Chapter 12
Maths Adds up

Maite Gorriz and Santi Vilches

Abstract In this chapter, talking from our professional experience in Secondary Mathematics Education in Catalonia, we will first describe the essential elements of working with interdisciplinary activities and knowing how to carry them out from an organizational point of view. Subsequently, we will describe in detail two examples in which good learning outcomes were achieved. Both examples have arisen from the necessity to attend to all the capacities of the students in low secondary school (12–14 years old) and to give them the possibility to develop their own creativity. Furthermore both examples have an extension activity for students in upper secondary school (16–18 years old) where it is possible to attend to their abstract capacity and connect all their knowledge appropriate to their age. Furthermore, we have chosen these two cases because both are very close to the reality of the students and both have an important background: to the desire for, and valuing of, a critical attitude. In the concluding section, we define some features of these interdisciplinary activities which can be extrapolated to any educational centre.

Keywords Interdisciplinary activities · Creative activities · Context · Assessment · Cooperative learning

12.1 Introduction: A New Approach to Teaching Mathematics

Mathematics has always been the spine of knowledge that has allowed every scientific, technological and artistic discipline to develop and grow. Let's think, as some examples, about the physical laws, the chemical formulations, the technological structures, the musical scales, the proportions in art, the philosophical logic or

M. Gorriz (✉) · S. Vilches
Vilamajor High School, Catalan Ministry of Education, Carrer de Can Llobera s.n., 08458 Sant Pere de Vilamajor, Barcelona, Spain
e-mail: mgorriz@xtec.cat

S. Vilches
e-mail: svilches@xtec.cat

B. Doig et al. (eds.), *Interdisciplinary Mathematics Education*,
ICME-13 Monographs, https://doi.org/10.1007/978-3-030-11066-6_12

the syntactic structures in linguistics. Let's even think about the harmony, beauty and emotional transmission. Mathematics has such a creative power that even if you wanted to escape from its theoretical hermeticism and try to make an abstract art piece, completely free and without following any method, you could still describe this with some theory about chaos or randomness.

Contrary to fact, mathematics has often been taught disconnected from reality, as something independent. In some periods, especially for example during the 70s when an obsessive Bourbakism made mathematical teaching totally abstract, it did not have any connection at all. Luckily, some contributions like the ones Freudenthal has made, have provided new pedagogical tendencies. Tendencies which are closer to an approach of mathematics for everybody.

Nowadays the reality of teaching is quite complex. On the one hand, we can still find many teachers who teach mathematics in an ideal school where calculation processes have a huge importance. On the other hand, in our experience, we find teachers who incorporate some research on their didactic activities and somehow improve the mathematical learning. These two models coexist under the tradition. This is possible because in our country teachers have the responsibility for their own teaching model (so, "Everyone thinks his or her way is best"), and in the social perception that the mathematics we learn at school has no use. A social perception which is enriched by the new technological reality in which any calculation, even symbolic ones, can be done by anyone using simple but powerful Android applications.

A proposal in order to help mathematics restore its role as the spine of knowledge is to create new didactic methodologies. Ones that take into account the new technological reality in which students use maths to develop their creative potential in any of the knowledge's disciplines. So as to reach this goal, we suggest the learning of mathematics using interdisciplinary activities as the starting point.

In the next section, we will present our experience of the interdisciplinary work in which we have been involved as part of our role as high school mathematics teachers. In particular, we will detail two examples that have been developed in different schools and have been taught from low secondary level through to upper secondary level. We started in interdisciplinary work because we needed to attend to different students' capacities and the first aspect to solve was language problems. So, we started working with language teachers in a wide context and the necessity naturally appeared to involve more subjects: science, arts, humanities, etc. Our examples give a flavour of the range of interdisciplinary work we engage with. And finally, the concluding section describes some of the main points of view that have been developed from this experience so far.

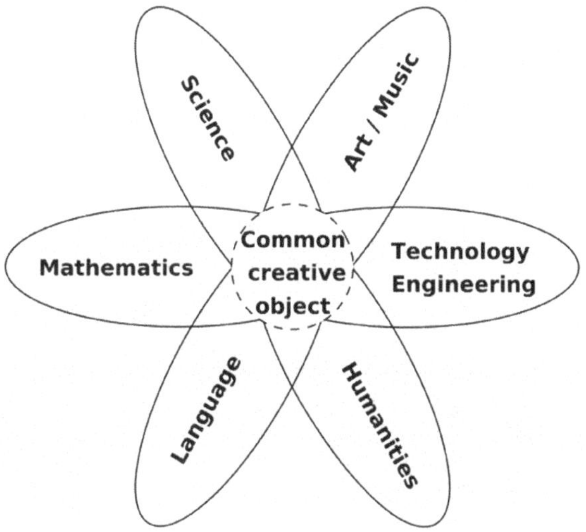

Fig. 12.1 Different specialists develop the educational curriculum around a final outcome

12.2 Interdisciplinary Activities: Form and Requirements

12.2.1 Form: Different Specialists Develop the Educational Curriculum Around a Final Outcome

Throughout our professional career, we have carried out interdisciplinary activities in different ways. We consider extremely important not only designing the interdisciplinary activity but also knowing how to put it in practice. So, we propose coordination between teachers based on the educational curriculum of each subject. The idea is to define a final outcome, a "common creative object" which has to be built (Miettinen, 1998), one for each student, and design a common working guideline for the students as the starting point to work on curricular content within every subject (in each class). That is, each specialist teacher works in his/her subject with a common guideline in order to guide each student to build a common creative object from his/her subject. For example, in the interdisciplinary activity "FX Thales", the common creative object is to create a short film with special effects. The subjects involved are Language, Technology and Mathematics. The language teacher works with the script, the technology teacher works with production of the short film and the maths teacher works on special effects using Thales Theorem (Figs. 12.1 and 12.2).

The advantages of this form are that the learning process is close to current social necessities where jobs often involve interdisciplinarity, team work and the need to improve. This kind of learning is achieved through the common working guideline for every subject, a cooperative group methodology and with an assessment as part of the activity. Furthermore, this form accepts that teachers do not necessarily need to be specialists in more than one subject. For instance, in Catalonia, a teacher only needs

Fig. 12.2 A frame of "We don't need to call a tow truck". An example of a short film. 13-years old interdisciplinary activity. INS Pla Marcell, 2017

one specialization in order to get into Secondary Education. It is obvious (to us) that a good teacher needs to be able to teach under a multidisciplinary perspective. However, we would argue, only a specialist is capable of properly structuring students' learning in a particular way, as Watson and Barton (2011) suggest. This is especially clear in mathematics. It is difficult for a non-mathematician to understand how a teenager's brain makes a step forward to abstraction. On many occasions, abstract or geometric procedures, like Thales theorem, are taught as a pure and rigid algorithm and there is a lack of concern in how a teenager perceives these procedures.

12.2.2 Requirements of Interdisciplinary Activities

Interdisciplinary activities promote teacher collaboration (St. Clair & Hougs, 1992). From our experience, to work with interdisciplinary activities, the starting point must be coordination between teachers of different subjects. In this initial coordination, two things are needed: a reflection about the curricular content that has to be taught in each subject and the defining of a single common outcome that each student has to achieve. "Common" meaning that this outcome is the same in every subject that is part of the activity. Experience shows that defining this common outcome helps specify the curricular content of each subject.

This initial coordination between teachers has to lead to a working guideline for students which will lead to the achievement of the outcome by each one of them.

This working guideline needs to be the same in every subject involved. However, each part of it will be done by the specialist teachers. This way, each teacher will be in charge of helping students learn essential topics of the corresponding subject within his or her own professional didactic criteria.

According to Van den Heuvel-Panhuizen and Drijvers (2014) "realistic situations are given a prominent position in the learning process". We propose that the common outcome must be defined really well and must be feasible and close to students, so that the achievement of the outcome is meaningful for them from the beginning. This guarantees students' motivation and their involvement in the learning process.

In Realistic Mathematics Education, Freudenthal (1973) argued "students should be active participants in the educational process". So, we propose that each student develops their own creative ability in accomplishing the outcome and constructing their own learning process. In order to reach this goal, students' task lies on writing a report describing the whole process. This way, the teaching and learning process adjusts to each student's abilities. And this is how we take care of student diversity.

It is essential to work in groups, cooperating and using heterogeneous teams. Although each student writes his or her own report about the process and creates his or her own final outcome as an individual work, a general methodology of cooperative work is suggested in order to enhance learning using role assignment. Every student will later analyse their classmates' outcomes, enriching their own knowledge by others' ideas. When these ideas are their own, or from their classmates, we find that motivation is highest and both interest and learning increase. And we assess the learner's creative work on open-ended innovative tasks and their capacity to work together (Howes, Kaneva, Swanson, & Williams, 2013).

The teacher role is that of a facilitator. Teachers should be trained to help in developing new ideas, adapting to each situation and finding tools to help students to achieve their goals. Technology should have a key role in facing and achieving new situations. For example, using photography or video to visualise new mathematics concepts or using functions or statistic structures in some specific software to solve higher level problems. So, teacher's knowledge about his or her subject and about didactics of his or her subject are the key to a proper orientation of the learning process of students.

From understandings of assessment for learning (Black, Harrison, Lee, Marshall, & Wiliam, 2004), assessment must be integrated within the learning process, answering to the following question: "How can I improve my creation?" It is not the outcome that matters most but the way to achieve it. This is why the aspects that need to be considered during the assessment are: the language of the reports from a formal point of view, the mathematical content that helps accomplishing the desired goal, and every curricular aspect from each subject involved. Constant research for improvement and error correction with teachers' help is essential for the assessment. An important part of the whole learning process will consequently be providing students with the necessary tools to evaluate and improve their professional productions in the future. Assessment means learning to improve and, therefore, learning to evaluate. It includes 'self-assessment' in the self-regulation process (Panadero & Alonso-Tapia, 2013).

The last requirement for interdisciplinary activity is coordination between the teachers involved in the process in order to do a final assessment of the interdisciplinary activity once it is finished.

We have outlined what in our experience are the most important aspects of interdisciplinary work in our context, we now turn to look in more detail at two cases focused on particular tasks which we have carried out in recent years. We have chosen the following cases because they are familiar to teenagers. When you see a group of teenagers in the street, they are drinking a soft-drink and listening to music. Aren't they? So, our task as teachers, would be to develop an activity around a "package of 33 cl, like a can" and around "music" from different subjects. Furthermore, another important aspect is that this activity should provide students with value and encourage a critical attitude. For example, why don't companies manufacture packages following ecological criteria in spite of commercial criteria? How are emotions involved in choosing a package or in listening to music?

12.3 Case Studies

12.3.1 Case Study 1: Create Your Own Package (12–14-Year-Olds)

This case study engaged students in designing a package for an object of unspecified shape but with a volume of 33 cl. The original idea was to study geometry while avoiding boring exercises of volume calculation from a book. We asked ourselves: how can students understand volume (3 dimension) with pictures (2 dimension) in an unreal situation? On the other hand, which volumes can be familiar to a teenager? The answer was "building a can". So, we decided to study the volume contents in the geometry curriculum on the basis of students' own packages of 33 cl. This way we can have as many "real volumes" as students in the class. After this first idea, we (maths teachers) saw that arts and language teachers were facing the same problem in their classes when dealing with the unfolding 3-dimensional figures and advertisement language respectively: they needed a real situation, too. So, this was the starting point: we needed to work together!

12.3.1.1 Context, Curriculum and Assessment

This interdisciplinary activity has been developed in the following centres during the specified periods of time and with different subjects involved (Table 12.1).

In each school, the activity is first organised with a meeting of the various subject teachers to decide the educational curriculum which they will work with. So, the educational curriculum could change every school year. Table 12.2 is an example from school year 2015–2016.

Where Physics has been involved, in Montserrat Roig High school and Marta Estrada High school, the educational curriculum has included Archimedes' principle, mass, density, force and pressure.

The common outcome, as already stated, is a package of 33 cl. Initially, this package had to be designed for soda. However, when Science became part of the interdisciplinary activity it was agreed that it only needed to be a package for some food product. In this task, the creative process of the students is of great importance, and each subject too.

Each subject determined some specific goals:

Maths: The mathematical challenge lies in calculating the volume of different figures and adjusting the measures to a set volume. In maths class, students will make the initial design, the necessary calculations, and will develop their geometrical competence.

Science: In this case the challenge lies in deciding which food item will be in the package and which nutritional information the tag has to include. To do so, students have to analyse some tags from different products and think about the consumption of some products taking the food pyramid into consideration. They will also have to study the energetic value of the food item they have chosen.

Physics: Here, Archimedes' principle is used so as to check the final volume of the object. Notions of mass, density, force and pressure of different food produce inside a package are also researched.

Table 12.3 Where, who, when and how are involved in the interdisciplinary task "Creating musical instruments"

Centre	Involved students	School year	Involved subjects
El Sui high School (Cardedeu, Barcelona)	15 students	2011–2012, 2012–2013	Maths (and Statistics), Technology, Music, Geography
Arquitecte Manel Raspall high School (Cardedeu, Barcelona)	96 students	2014–2015	Maths (and Statistics), Technology, Music
Marta Estrada high School (Granollers, Barcelona)	60 students	2016–2017	Maths (and Statistics), Music, Physical Education, Linguistics (Catalan and English), Arts
Vilamajor high School (Sant Pere de Vilamajor, Barcelona)	96 students	2017–2018	Maths (and Statistics), Music, Arts, Technology and History

Table 12.1 Where, who, when and how are involved in the interdisciplinary task "Create your own package"

Centre	Students involved	School year	Subjects involved
El Sui High School (Cardedeu, Barcelona)	92–96 students	2010–2011,…, 2016–2017	Maths, Art, Linguistics
Montserrat Roig High School (Sant Andreu de la Barca, Barcelona)	110 students	2012–2013,…, 2016–2017	Maths, Art, Linguistics, Physics
La Sínia High School (Parets del Vallès, Barcelona)	60 students	2013–2014,…, 2016–2017	Maths, Art, Science
Arquitecte Manel Raspall High School (Cardedeu, Barcelona)	96 students	2015–2016 2016–2017	Maths, Art, Linguistics, Science, Technology, Music
Marta Estrada High School (Granollers, Barcelona)	60 students	2016–2017	Maths, Art, Linguistics, Science, Physics
Pla Marcell High School (Cardedeu, Barcelona)	95 students	2016–2017	Maths, Linguistics, Technology

Table 12.2 Educational curriculum worked in Arquitecte Manuel Raspall High school, 2015–2016

Subject	Educational curriculum
Mathematics	Volumes. Unfolding figures
Art	Tag design, technical drawing, plane representation systems
Linguistics	Advertising vocabulary
Science	Food pyramid
Technology	Materials
Music	Music and society

Art — Students will make the artistic design of the tag. In some grades the unfolding 3-dimensional figures have been studied too.

Technology — The challenge lies in studying the possible materials with which the package could be made, considering their features and deciding which one they will use. The package will be created at the technology laboratory.

Linguistics — In this case the challenge lies in making an advertisement to promote the product they have created.

Music — Adding music to the advertisement.

Fig. 12.3 Heterogeneous working groups

Teachers then create the working guideline with the activities from their subject, following the agreed common cooperative work guidelines. It is highly important to keep in mind that the interdisciplinary activity "Create your own package" is not done after the subject activities, for example, on volume in a mathematics class. It is itself the activity on volumes (or the linguistic activity on advertising language, etc.). The guideline contains everything necessary to carry out this interdisciplinary activity. The same happens for each subject involved.

Cooperative work guidelines define that each heterogeneous group of 4 or 5 students represents a team from a fictitious food company in charge of creating a new product line. They will have to decide which product to put in the package (Science) and design a corporative logo to add to this one (Art). They can choose or not choose homogeneous shapes within the team members (Mathematics). They will have to present their commercials together (Linguistics and Music) in order to promote the new product line. They will also have to help each other in cooperative work but, even though they team up, each student must contribute with their own individual task in creating their own package (Fig. 12.3).

At the beginning of the interdisciplinary activity, teachers from each subject explain the activity, the challenges students need to overcome in that subject in relation to the final object and the assessment criteria are shared by students and teacher. Each subject is studied using the one and only working guideline during the assigned hours within the usual schedule. In every case, the interdisciplinary activity "Create your own package" was carried out during the same weeks in each subject involved.

Students are asked to produce a report for each of the subjects as part of their learning process assessment. Each subject gives the guidelines to fulfil the report within the working guideline. The working guideline explains how to write the report about each student's process so that it can become a tool of assessment. This tool allows students to modify their learning process so as to improve it. The final object and the chosen procedure will also be taken into account according to each student's skills. This assessment will be used in order to motivate students in a more personal way. Last but not least, calculations about others' packages and a group reflection about the new product line created will also be part of the assessment. Each one of these items will be taken into account to assess students' acquired knowledge. This whole process and reflection will enable also a teachers' assessment and students' grades analysing every possible item which can be improved in their learning process. Grades will be agreed with students themselves individually and will be used to inform the families about their progress. It will be considered as another learning item for students.

12.3.1.2 The Activity in Mathematics

In Mathematics, the teacher observes, reflects and takes part in assisting the students to make steps toward abstraction. The teachers will guide students so that they are able to create their package according to their own abilities. Therefore, we can have some students working with simple prisms (just a cube, for example) and some others designing highly complex shapes. Regarding the volume calculations for the package, some students can work with small shapes to fill in the cube using the Voxcad program (see Fig. 12.4).

Other students use the Archimedes principle to make sure their calculations are correct in the laboratory. Others do calculations using abstract formulae. And some use the dynamic geometry software GeoGebra. Dynamic geometry allows students to have a different vision of Maths and helps enormously in developing their creative processes (see Fig. 12.5).

Moreover, teachers help students to develop their creative abilities in asking them to freely create the new packaging. This "freedom principle" regarding work is seen in every subject (advertising, tag design…) but it is important for students to realise how big the creative power of Mathematics is. The only restriction in the task is the final volume the package must have. This restriction makes students find connections between measures, shapes and volume (Fig. 12.6).

Each student is asked to calculate the volume of every package created by their classmates in order to practise and consolidate different techniques and formula about volume calculation. This way they won't be working with exercises detached from reality or with exercises that can be found in any book. They will be working with activities based on their classmates' productions and which consists of analysing a physical and tangible object. This helps them to notice and reflect on any mistakes that they make in the calculations.

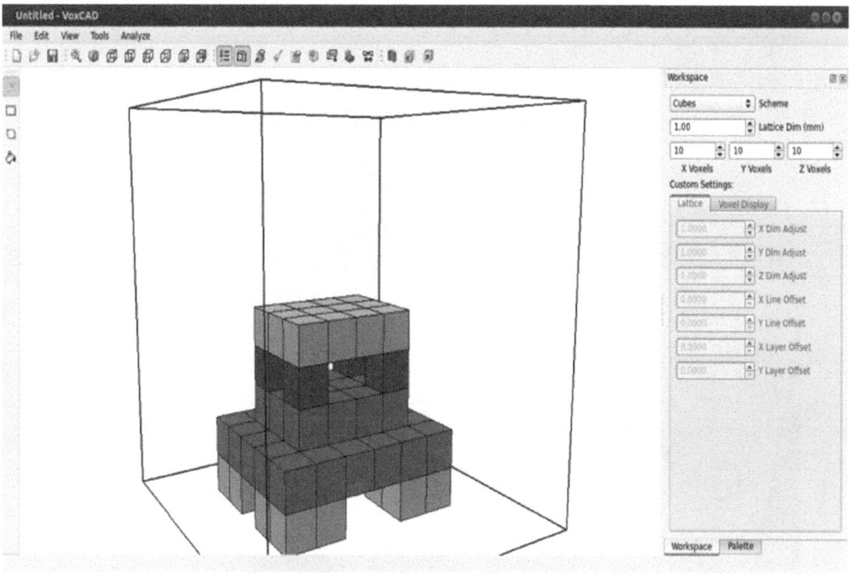

Fig. 12.4 A student's voxcad solution

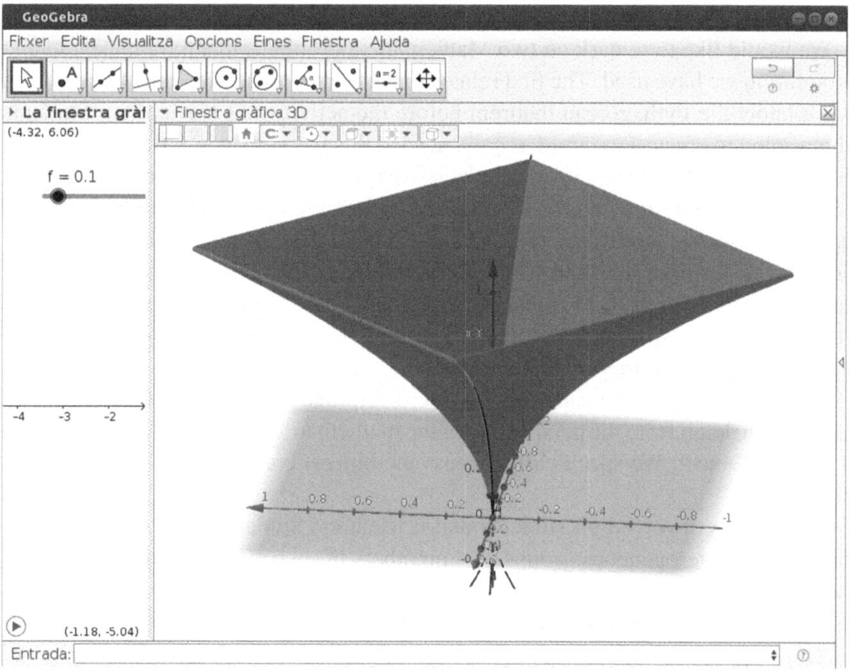

Fig. 12.5 A student's GeoGebra research

Fig. 12.6 A student's creative object

We would like to remark on two Mathematical problems that have appeared and the solution we have used. The first relates to the Pythagorean theorem. The students didn't know the Pythagorean theorem before the activity was carried out. Students who wanted to create a pyramid or cone needed this theorem in order to know which measures of the flat design of the package were required. This problem was solved using GeoGebra or drawing the triangle and then measuring its unknown sides. The second problem relates to irregular shapes. Some students created packages with a totally irregular base. In order to solve this problem, the "covering with spaghetti" technique was used so as to calculate the base's area. This consists of covering the whole shape with spaghetti and then adding up each piece of spaghetti's area (see Fig. 12.7). It is an approximate integral technique.

Here we explain the results of the interdisciplinary activity "Create your own package" in detail from the perspective of the mathematical complexity and difficulty of the final object. We specifically discuss the objects created at our own school in 2015–2016.

The activity involved 96 students and 6 teachers. Students were divided into 4 classes with heterogeneous groups of 4 members in each. 10.5% of the 96 students had learning difficulties.

Students created their packages with the following results:

(i) 3% of the students didn't create any package due to serious discipline or absenteeism problems. These students didn't have any learning difficulties.

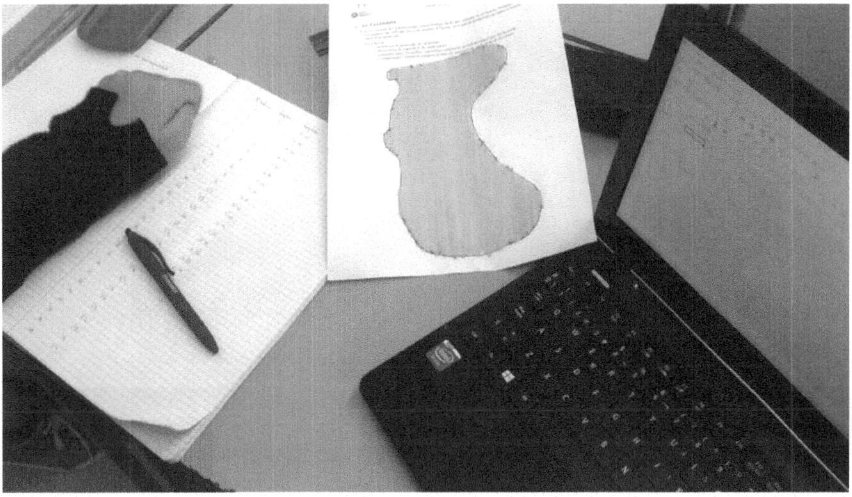

Fig. 12.7 Using an approximate integral technique

(ii) 42% of the students created their packages using a basic shape. In these cases, the volume calculation consisted of calculating the base's area and multiplying it by its height (including cylinders or prisms with a rectangular or polygonal base).
(iii) 22% of the students created packages formed by two or more basic shapes such as a cube and a cylinder.
(iv) 26% of students created figures which required a complex calculation of its volume or construction. This included, for example, pyramids, cone's trunks, spheres or semi-spheres.
(v) The final 7% constructed completely irregular figures which required a highly difficult volume calculation. Considering the students' ages and knowledge, this meant only making an approximation, using tools such as GeoGebra (Fig. 12.8).

12.3.1.3 Impact of the Activity

In order to evaluate the students' mathematical knowledge, they also had to complete a test in which they were asked to calculate the volume of a given three-dimensional geometrical figure. These were the results:

(i) 7% of the students were not able to solve the problem.
(ii) 15% made some serious mistake.
(iii) 25% had some difficulty or made a small mistake.
(iv) 52% calculated the volume correctly.

Fig. 12.8 Examples of student packages (top left, basic; top center, mixed; top right, complex calculation, bottom, irregular figure)

Elsewhere, in Assessment Board of the Educational System[1] for example, there are some indications that students at the schools which engage in these interdisciplinary activities do better than average in Catalan standardized tests at age 16, but this requires further research to substantiate.

In general, teachers assessed the overall experience in a positive way. Teachers said that students had a higher involvement when they worked with their own creative production and with "real objects". For instance, a comment by a science teacher in Arquitecte Manel Raspall High school, is very illustrative. She wrote in her final assessment report of the activity "Create your own package" 2015–2016: "I don't really know if this activity improves students' results. But what I do know is that it is an activity which gives the opportunity to connect many things and this fact makes students think and reflect, which is always a good thing".

In relation to the coordination in an interdiciplinary activity, teachers felt it was useful knowing what teachers from other subjects were doing, because they could then make teaching much more rational through finding out common knowledge. It was also seen to help avoiding unnecessary repetitions and having different approaches to the same content. However, the majority of teachers involved men-

[1] http://csda.gencat.cat/ca/arees_d_actuacio/avaluacions-consell/.

tioned that the activity demanded a large amount of meeting hours with each of the other teachers involved, and they did not always have availability for this. This was probably the biggest concern when carrying out this interdisciplinary activity.

However, from the assessment point of view, teachers agree easily about suggestions for improvement in the formative assessment of each student and their marks because methodology are similar and there are some common assessment criteria in each subject involved in the interdisciplinary activity.

12.3.1.4 Continuing the Activity

"Create your own package" is an interdisciplinary activity for 12–14 year-old students but we have extended this activity from a different point of view for 16–18 year-old mathematics students. In this case, it is not an interdisciplinary activity, as there are not more subjects involved, but there are connections with physics, technology and social aspects. It is really interesting to design packages with students who know derivative functions and have the tools to work with the curricular content of integrals. We work with the design of a one-litre-capacity package (bottles, jars, decanters, etc.) using the dynamic geography program GeoGebra 3D and the modification of different parameters of polynomial functions as the starting point. In fact, the activity consists only in designing a package and seeing it in a virtual environment. It's impossible to build the outcome till now. However, in Pla Marcell High school we are working with engineering teachers because they have bought a 3D-printer. So, we are preparing a new interdisciplinary activity together for the next course. We will see!

The most important thing for us is to continue this interdisciplinary activity in each of the schools where it has been put in practice, although subjects may vary depending on the school year.

12.3.2 Case Study 2: Creating Musical Instruments (11–13-Year-Olds)

This case study engaged students in creating musical instruments. The original idea was to study feelings which are produced by music using mathematical or statistical methods. Music teachers listened to all kind of music with teenagers and it was a good activity because students were completely hooked. So, music teachers wanted to do some activity more deeply. On the other hand, maths teachers wanted to join mathematics and music in a single subject in order to study progressions, integer numbers, powers and roots, functions, especially logarithmic and exponential ones, and statistics. So, maths teachers and music teachers started to work together but we needed to define which feelings could be involved. This way, language teachers joined us. And finally, could it be possible to perform a concert where students played their own instruments?

Table 12.4 Educational curriculum put into practice in Marta Estrada High school, 2016–2017

Subject	Educational curriculum
Mathematics	Progressions, integer numbers, power and roots Statistics: average and diagrams
Music	Music and feelings
Physical Education	Mime
Linguistics	Objective and subjective text. Vocabulary about feelings Descriptive text. Idioms
Arts	Colours and feelings

12.3.2.1 Context, Curriculum and Assessment

This interdisciplinary activity has been developed in the centres during detailed periods of time and with different subjects involved (Table 12.3).

As in the previous case, in each school, the activity began with a meeting which brought together teachers of different subjects, in order to decide the educational curriculum they would develop. Table 12.4 is an example from school year 2016–2017.

When Technology has been involved, in El Sui High school and Arquitecte Manuel Raspall High school, the educational curriculum included materials and tools. Also, Geography was studied at El Sui High School using different instruments and musical scales from different cultures as the starting point.

The common outcome here is a performance where students play instruments created by themselves, while communicating feelings. In order to achieve the common outcome, students have to aim for the following specific goals in each subject:

Maths — The mathematical challenge lies in calculating progressions and distinguishing arithmetic ones from geometric ones. In arithmetic progressions students go deeper into integer numbers and in geometric progressions students go deeper into powers and roots. After that, students argue about the musical scale and the relation between notes and size and shapes of “musical instruments”.

Statistics — The statistical challenge lies in using diagrams in order to assess feelings. Each student experimentally measured the necessary length of a tube so that it could generate a specific frequency (using a digital instrument tuner). Studying the average will give us the chance to experimentally determine each tub’s length and prove that it adjusts to a geometrical progression. Statistical parameters which enable us to define which emotion a specific music creates are also studied. We work using spider diagrams and the averages came from the votes of the listeners to the music.

Music — In this case the challenge lies in listening to different kinds of music and deciding which feeling will be involved. While studying how

Fig. 12.9 Preparing their performance playing their own musical instruments. Marta Estrada High School, 2016–2017

	to listen to music, emotions (defined at language class) awakened by different kinds of music were analysed (using statistical tools). Acoustic foundations of the instruments students made were studied and musical interpretations with these instruments were prepared.
Physical Edu	Mime is the tool to identify the different traits of the transmission of feelings through the expression.
Language	The challenge lies in defining feelings and producing text expressing them by idioms, poetry, etc.
Art	Students will make the artistic work with colours and feelings.
History	The importance of the music in the ancient history: culture, religion and tradition.

Teachers then create the working guideline with the activities from their subject and define a cooperative work where each heterogeneous group of 4 or 5 students have to choose a feeling (Linguistics) and prepare their performance (Physical Education and Art) playing (Music) their musical instruments (Maths) about that feeling. At Marta Estrada High school, the musical instruments have been built in Maths sessions because the Technology teacher has not been involved.

Pan flutes made of PVC tubes and idiophones made of cardboard tubes were created in El Sui High School. In Arquitecte Manuel Raspall High School they also made tambourines, rain sticks and drums. And in Marta Estrada High school they also made sophisticated idiophones (Fig. 12.9) and maracas (yellow object in Fig. 12.9).

Assessment is totally integrated within the learning process due to the fact that if calculations were wrong, the instruments would not sound properly. Furthermore, students assess the performance of each other group of students using statistical diagrams in order to evaluate feelings. Finally, they share opinions about the performance and feelings.

12.3.2.2 The Activity in Mathematics

In Mathematics, the first step is to study progressions and relate them to musical scale. It's easy to start with arithmetic progressions. From a curricular point of view, Maths teachers were interested in connecting this activity to order of operations with integers. So, for instance, a detailed study of the numerical values of different successions made it possible to contextually justify multiplication rules. On the one hand, "doing" was connected to the positive sign and "undoing" to the negative one. On the other hand, going up was positive and going down was negative. In this way, in a numerical sequence, undoing a down meant going up. Which means that negative (undoing) of a negative (down) is positive (going up).

Then, we continue with geometric progressions and the measures of the "tubes". In particular, one of the goals of this activity was to experimentally check if the measure of the pan flute's tube had to be a geometrical progression of common ratio twelfth root of 2 (well-tempered scale). Surprisingly, it is like this with the big cardboard tubes used to create the idiophones but not with the small PVC tubes used to create pan flutes. The lack of a physics specialist made it impossible to analyse the reason why. However, an activity about measuring the necessary length of a cello string to create a tempered scale was added. The presence of the geometrical progression of common ratio twelfth root of 2 was verified with this experiment. This subject became an interesting discussion with the students and created a big interest in doing experiments and numerical testing.

At the same time, students calculate averages when they are needed. For example, at Arquitecte Manuel Raspall High school each student made their own calculations about the length of the tub to make the pan flute and shared the results. Average was calculated and 5% of the values were far away from the average so they had an implied mistake. These values were checked or rejected. Furthermore, spider diagrams are studied in order to evaluate feelings. Those diagrams will be used in Music class and they will be the tool to assess their colleagues' performances (Fig. 12.10).

12.3.2.3 Impact of the Activity

At Arquitecte Manuel Raspall High school, in later internal tests meant to detect the routine knowledge about integers operations, results were surprisingly good. More than 80% of the students were detected to properly solve this kind of exercise. While in other courses where this interdisciplinary activity wasn't carried out, the level of achievement was under 50%. However, this requires further research to substantiate.

Besides, the use of spider diagrams to do the emotions study was easily achieved by students. Later on they used them to make reflections on how to improve different kinds of self-assessment in tutorial sessions and in other subjects like social sciences.

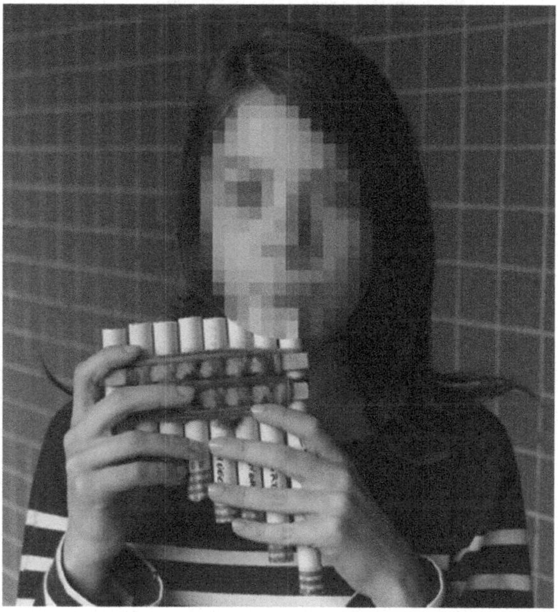

Fig. 12.10 Playing her pan flute. Arquitecte Manuel Raspall High school, 2014–2015

At Arquitecte Manuel Raspall High school, the assessment made by teachers was highly positive, especially for the technology and music teachers. On the one hand, technology teachers were able to work the "technological process" part of the curriculum with a final outcome that was more than just the typical object that students bring home to decorate. On the other hand, music teachers felt more valued since students showed more interest in the music subject than usual. It is also important to mention the mathematics teachers' satisfaction when improving integer calculation as a consequence of this interdisciplinary activity. At Marta Estrada High school, in a coordination meeting, language teachers (Catalan and English) said that it has been very useful to simultaneously study linguistic aspects like definitions, idioms and poetry because students have compared both languages spontaneously. At Vilamajor High School, like in Marta Estrada High School, the most important thing in the assessment by teachers (and students) was the final concert using their own instruments. Furthermore, in Vilamajor was a public performance playing flutes and singing while showing the process of construction in the technology lab (Fig. 12.11).

12.3.2.4 Continuing the Activity

"Creating musical instruments" is an interdiciplinary activity for 11–13 year-old students and we have continued with this activity from a different point of view for 16–18 year-old mathematics students. In that case, the math teacher is the only teacher

Fig. 12.11 Final outcome: A public performance of music using their own flutes and showing their construction in technology lab, 2018

involved because in Upper secondary education curricula, Music is not included in the Technological branch. The same way Maths is not included in Arts branch.

Students analyse the musical scale, which sequence it belongs too, and they find out the exponential function. Then the Maths teacher asks students which music they could make if instead of using the exponential function they used other functions (like the parabola, trigonometric functions, …). In this activity, students must find the hypothetical music frequencies that different functions would create and program them using PureData, a software program that gives them the option to listen to music played with a freely programmed musical scale. This interdisciplinary activity makes students think about the features of every basic function using such a ludic and dynamic aspect as musical interpretation is.

Finally, another activity about musical scales is "The Vicenzo Galilei rebellion: when music was part of mathematics". It has been carried out in the university for the last three years, as part of a mathematical talent development program called "Crazy for Mathematics", targeted at 16–17 year-olds and it is guided by the authors. This activity presents to the students different musical scales along history from a Maths point of view. Then, students discover that it is impossible to have perfect musical scales mathematically because if we accept the harmonics, we will have an imperfect interval. In the opposite way, if we have perfect intervals, we will not respect the harmonics. So we, as Math teachers, propose to students to create a new scale using mathematical criteria. Finally, students have to perform a composition using their

Fig. 12.12 Showing their musical scale research by Pure Data in "The Vicenzo Galilei rebellion: when music was part of mathematics", 2016

own scale (Fig. 12.12). In the last edition of this activity, on February 2017, students' assessment of this activity averaged 9.4 out of 10.

12.4 Conclusions

In this final section we would like to present the conclusions from the experience of our interdisciplinary activities', including those described above, in which we have been engaged with over the last seven school years, both directly as teachers working with our own students, and as trainers of other teachers who have carried them out at their own schools.

First of all, interdisciplinary activities promote a learning improvement in students, not only in the mathematical skills but also in other key competencies. (We believe that they also improve skills in other subjects even though we cannot prove it). It is not possible to establish a direct connection with the improvement in national tests though since many factors have an influence.

In Mathematics, interdisciplinary activities allow us to give a new role to mathematical teaching, which promotes creative abilities in students and makes mathematics more than just a calculating tool. Notably, every student takes part in his or her own active learning process due to several reasons. Firstly, because the interdisciplinary activity offers to students a consistency between the subjects which are involved.

Furthermore, the activity is about a reality close to students (cans, music, feelings, etc.) and therefore it engages students. Secondly, because collaborative work forces students to be part of the group since the other members themselves make them work. And finally, because they see the task as suitable to their cognitive and creative abilities. On the other hand, this kind of activity includes an assessment that ensures students are active from the beginning with producing the report and achieving the final outcome.

Co-ordinated interdisciplinary activity allows teachers to know what teachers from other subjects are doing and helps them make teaching much more rational in finding out common knowledge. It also helps avoiding unnecessary repetitions and having different approaches to one and the same content. However, the activity demands a big amount of meeting hours with each of the other teachers involved, and they do not always have availability for this. This is probably the biggest concern when carrying out this interdisciplinary activity. Another difficulty is how to enrol teachers to work within an interdiciplinary activity. From our experience, it's a "seduction problem": we have always collaborated with friendly teachers.

Personally, we have enjoyed a lot working with interdisciplinary activities because students are more cheerful. So, their learning improves not only in Maths but also in other subjects involved. Furthermore, we have gained from interdisciplinary activities because we have taught collaboratively with teachers from other subjects and, as a result, we have improved our teaching.

12.5 We Encourage You to Try It

Three variables constitute the instructional core (City, Elmore, Fiarman, & Teitel, 2009), that is the teacher, the learner and the contents. These variables are interdependent and in constant motion. Thus, a change in any of these variables has an impact on others and on the final result. A student's academic success results from the quality and frequency of the interconnections between student, teacher and content (Salavert, 2015). So, a competent teacher should achieve each student to be aware of four aspects about their own learning: what should they learn, why should they learn that, how should they do it and how should they improve it. Students should achieve the outcome in their learning process if teachers propose them a creative, closer and motivating challenge. In conclusion, students lead their own learning because they know the purpose of their learning through competence activities (Gorriz & Nuñez, 2013).

In this chapter, we have proposed these activities which encourage students to reflect on their progress by sharing what they have learned and by asking questions about what they did not understand and teachers use this ongoing formative feedback to inform their guidance in a continuous, differentiated and enriching loop. Furthermore, these activities allow to show knowledge as interdisciplinary from Maths. Teachers from different subjects find a common final object to develop students learning. All of us think that "it's better to work with the same activities in a different way (from students creativity) rather than to work with different activities in the same way" because students motivation is essential.

These activities allow both the curriculum and teaching to develop around the learners' own interests and these emphasise students engagement in their formative assessment. These proposal is coherent with the new project of law about assessment in Catalonia which is according to the Competence based Curriculum (decree 187/2015). However, Assessment Decree project would had been published a few month ago but he Spanish government has paralysed the regional government of Catalonia from October 2017.

From July 2017, Innovation and Training department of Ministry of Education of Catalonia is promoting a training program about competence assessment (formative assessment and selfassessment) and we, the authors of this chapter, are coordinating the training in Maths as a subject. In this training program, we promote interdisciplinary activities and encourage Maths teachers to find partnerships in other subjects in order to encourage learners to take control of and regulate their own learning in a creative environment.

References

Black, P., Harrison, C., Lee, C., Marshall, B., & Wiliam, D. (2004). Working inside the black box: Assessment for learning in the classroom. *Phi Delta Kappan, 86,* 8–21.

City, E., Elmore, R., Fiarman, S., & Teitel, L. (2009). *Instructional rounds in education: A network approach to improving teaching and learning*. Cambridge, MA: Harvard Education Press.

Freudenthal, H. (1973). *Mathematics as an educational task*. Dordrecht: Reidel Publishing.

Gorriz, M. & Nuñez, J. (2013). *Lideratge per a l'aprenentatge: estudis de cas a Catalunya* (pp. 68–82). Barcelona: Fundació Bofill. http://www.fbofill.cat/intra/fbofill/documents/publicacions/581.pdf. Accessed 30 April 2018.

Howes, A., Kaneva, D., Swanson, D., & Williams, J. (2013). *Re-envisioning STEM education: Curriculum, assessment and integrated, interdisciplinary studies*. A report for the Royal Society. https://royalsociety.org/%2A/media/education/policy/vision/reports/ev-2-vision-research-report-20140624.pdf. Accessed 30 April 2018.

Miettinen, R. (1998). Object construction and networks in research work: The case of research on cellulose-degrading enzymes. *Social Studies of Science, 28,* 423–463.

Panadero, E. & Alonso-Tapia, J. (2013). Self-assessment: Theoretical and practical connotations. When it happens, how is it acquired and what to do to develop it in our students. *Electronic Journal of Research in Educational Psychology, 11*(2), 551–576. http://dx.doi.org/10.14204/ejrep.30.12200. Accessed 30 April 2018.

Salavert, R. (2015). Coaching: An apprenticeship approach for the 21st century. *International Journal of Educational Leadership and Management, 3*(1) (2015). https://dx.doi.org/10.4471/ijelm.2015.02 Accessed 30 April 2018.

St. Clair, B. & Hough, D. L. (1992). Interdisciplinary teaching: A review of the literature. http://files.eric.ed.gov/fulltext/ED373056.pdf. Accessed 30 April 2018.

Van den Heuvel-Panhuizen, M., & Drijvers, P. (2014). Realistic mathematics education. In S. Lerman (Ed.), *Encyclopedia of mathematics education* (pp. 521–525). London: Springer.

Watson, A. & Barton, B. (2011). Teaching mathematics as the contextual application of mathematical modes of enquiry. In: T. Rowland & K. Ruthven (Eds.), *Mathematical knowledge in teaching* (pp. 65–82). Dordrecht: Springer.

Chapter 13
The Successful Students STEM Project: A Medium Scale Case Study

Linda Hobbs, Brian Doig and Barry Plant

Abstract Schools in Australia and internationally are responding to calls to offer new and innovative learning opportunities in STEM. STEM stands for Science, Mathematics, Engineering and Mathematics, but when amalgamated into the acronym 'STEM' can potentially mean more than the sum of the four parts. In choosing how to respond to the STEM 'push', schools must first navigate through the many 'versions' of STEM emerging within the education community, and then face the task of upskilling teachers in content and language, teaching approaches, and new technology and equipment. Professional development of teachers plays an important part in assisting teachers and schools in this period of innovation and change. This chapter describes one such professional development project where teachers from ten schools in regional Victoria, Australia, were supported in developing new knowledge, language, pedagogy, and curriculum to support their development of a 'STEM vision' for their schools. The activities developed by these schools are outlined to illustrate that they each have taken a different approach to STEM, with case studies showing how these activities were developed. The factors critical to the success of the program are outlined, which have implications for a policy response, as well as challenges that may threaten the sustainability of such initiatives.

Keywords STEM · Professional development · Teacher learning · Interdisciplinarity

13.1 Introduction

The Successful Students STEM program (SSSP) is a medium scale professional development program based in a regional city in Victoria, Australia. Schools in the region have a relatively low rate of student enrolment in the senior STEM disciplines

L. Hobbs (✉) · B. Plant
Deakin University, Waurn Ponds Campus, Australia
e-mail: l.hobbs@deakin.edu.au

B. Doig
Deakin University, Burwood Campus, Australia

B. Doig et al. (eds.), *Interdisciplinary Mathematics Education*,
ICME-13 Monographs, https://doi.org/10.1007/978-3-030-11066-6_13

in comparison to the rest of the state, a fact that concerns both the schools and local industries.

Unlike many STEM case studies (e.g., English & King, 2015), this project involves a number of secondary schools, but more importantly, does not prescribe a single, 'one size fits all' model of STEM. Individual schools determine their own needs, and create their own perspective of what STEM can be in their school context. This has led to a range of STEM-focussed projects, ranging from Year 7 students from an all girls' secondary school designing and building a ramp for wheel-chair access to a part of their school, to another school designing and making vehicles.

SSSP offers schools assistance in preparing teachers for such STEM activities. Professional learning (PL) and ongoing supports for teachers are high priorities, as is the induction of school administrators into the concept of STEM and the importance of their role in creating a successful STEM program in the school.

The program is developed and implemented by a team of researchers from Deakin University. Funding supports teacher professional development, staffing and administration, support for partner schools to run school programs (such as excursions), teacher support for attending professional development and curriculum development, support for schools developing links with industry, quarterly STEM teacher network meetings (open for all STEM-related teachers in the region) and a national STEM education conference (http://stemedcon.deakin.edu.au/) where the SSSP teachers were showcased in a number of ways.

13.2 The Program

The programme involves 3 teachers from 10 partner schools committing to professional development for two years. These teachers may be Mathematics, Science or Technology teachers, or teachers in positions of leadership who can support the change process of the teachers and within the school generally. The professional development programme includes four professional learning (PL1-4) sequences of three phases each—Immersion (2-day intensive), Implementation (10–12 weeks at their school) and Review (1 day). Each sequence builds on the learning of the earlier sequences, with the foci being: PL1—Pedagogies and Contemporary STEM practices, PL2—Assessment and up-scaling to lead change, PL3—Sustaining change, and PL4—Embedding practice and generating evidence of change. The professional learning sequences are designed to progressively build teachers' capacity to plan, implement, evaluate and lead STEM teaching and learning in their schools.

This chapter reports on STEM initiatives from the first three PL cycles. These initiatives have thus far been units of work, learning sequences, or programming structures that incorporate some of the STEM practices and pedagogies that are promoted through the PL intensives. In addition, a Project Officer works with schools to support their developing practice. These PL sequences and the continuing support are key to supporting each school's approach to STEM innovation.

13.3 Negotiating the University-School Partnerships

The focus and structure of the professional learning was negotiated between the SSSP team and schools, and then re-negotiated to ensure the programme continues to meet the schools' needs. These negotiations are fundamental to the success of the programme as they ensure the Deakin team is sensitive to the ongoing and changing needs of the school. There is opportunity for feedback of data (particularly student attitude and aspiration survey data) to support school decision-making, and relationships are strengthened to support ongoing trust and reciprocity between the SSSP team, the teachers and school leaders.

13.4 The "STEM Vision Framework"

A framework called the "STEM Vision framework" (reported in Hobbs, Cripps-Clark, & Plant, 2018) was introduced in PL2 to provide a common language and direction for teacher decision-making about: conceptualization of STEM for the school; STEM pedagogies and practices; curriculum structure and teacher collaboration models; teacher learning and up-skilling; and using industry links. Development of a common language around these aspects of the STEM vision has been shown through an independent evaluation to be empowering for teachers for describing and guiding their practice.

Schools have been innovative in planning, implementing and sharing a range of initiatives that are at the forefront of thinking about new directions for STEM education. A survey of teachers after PL3 has shown that there has been substantial change from the teachers' perspectives: 50% reported evidence of embedded classroom innovations at 18 months and 75% at 24 months; 79% reported improved knowledge and understanding at 6 months; and 100% reported improved capability at 12 months. Evidence from the reporting days show that teachers are continuing to build their language around STEM, and the 'STEM practices' promoted through the programme is becoming part of their natural discourse around STEM. The STEM pedagogies and practices that have been introduced to teachers are listed in Table 13.1.

Tables 13.2 and 13.3 show the variety of projects that are emerging after three of the four professional learning sequences and the teacher collaborations, that is, how teachers work together to incorporate the different STEM subjects. Table 13.2 shows the teaching teams that are largely single subject-oriented in how they were implementing STEM, while Table 13.3 shows where subject integration was a key focus. In both tables, column 2 indicates the variety of curriculum areas that are being recruited to STEM; the majority are the Mathematics and Science teachers, but also included in some schools are the Information Technology and Materials Technology teachers, as well as the Arts teacher in School G in the third iteration.

Table 13.2 shows that Schools B, C and E were predominantly focused on improving the activities and learning outcomes in single subjects. In the absence of the

Table 13.1 STEM practices and teaching and learning practices

Interconnecting STEM skills/proficiencies[a]	STEM teaching and learning practices, where students…
Flexible reasoning skills	Problem solve Create Generate own questions Inquire
Effective and adaptable use of tools, processes and ideas (artefacts)	Use conceptual, digital, physical tools Explore and investigate artefacts Use a range of modern tools Use artefacts of the discipline in a flexible way Apply constructed artefacts to new contexts
Proficiency in professional/technical discourse	Understand and engage with the disciplinary representations Know the language Share and communicate Work in teams
Understanding of the nature of evidence in different settings	Collect real data in a variety of situations Use evidence to validate a solution to a problem or justify a decision Make judgements about the accuracy and reliability of information

[a]Adapted from Clarke (2015)

imperative to attend to the curriculum standards in multiple subjects, their projects were considered STEM-like because of their explicit link to the STEM practices through assessment that foregrounded these practices.

The other two schools, A and J, were delivered by teachers from one subject, but there were attempts to draw other subjects into the learning. School A projects were delivered by the Technology teacher who used the Science context as a way to engage students in the design process. School J involved the Science teachers (who also taught Mathematics) bringing the Engineering principles into Science classes. Similarly, the projects of School B involved some of the concepts from Science, but the projects were largely focused on re-designing the Mathematics curriculum.

In Table 13.3, there were different types of integration present. At School G a multi-disciplinary teaching team meant that parts of one project were delivered by different teachers in different subjects. In Schools F, H and I, the integration of subjects was done by each teacher, where one teacher makes explicit links or draws in ideas and learning outcomes associated with more than one subject, usually Science and Mathematics, but also Science and Technology, or all three in the case of School F.

While some teacher teams are predominantly Science (such as School E) or Mathematics (School B), the majority are inter-disciplinary; however, by the third iteration, even the subject-oriented teams were making links to their other subject teaching as most of the teachers teach both Mathematics and Science.

Table 13.2 SSSP projects per school in professional learning sequences 1–3 (PL1-3)—single subject-oriented teaching teams

Sch'l	PL#: year level (curriculum content)	Teacher collaboration	In these projects, students…
A	PL1 and PL2: year 8 technology (design processes and simple machines)	Tech teacher complements Science programme	Investigate types and uses of simple machines Design a machine that lifts 250 g weight
B	PL1: year 7 mathematics (angles, generating and using data), science (gravity and ramp slope)	Maths teacher makes connections to science and STEM practices	Investigate ramps for people in wheel chairs, the big question *Are all ramps the same?*: explore ramps in the community, test effects of angles on ramps, and design a school ramp
	PL2: year 8 mathematics (area, volumes, measurement and managing data), Science (transpiration and evaporation)	Maths teachers makes connections to science and STEM practices	Investigate *Are our garden water storage tanks large enough?*: analyse rainfall data, garden bed water losses, and size of water tanks in relation to the area of collection roofs
	PL3: year 7 mathematics (using data), science (ourselves)	Maths teacher makes connections to science and STEM practices	Investigate *When during my schooling do my parents spend most money on clothes?*: collect height data from the entire school study body and from local primary school children, and use secondary data to examine height changes with age
C	PL1 and PL2: year 7 info technology (developing coding skills)	Info Tech teacher makes connections to STEM practices	Complete online coding skill-building activities, leading to control of robotic lego devices
	PL3: year 7 and 8 info technology within science (coding and robotic control)	Science teacher makes connections to STEM practices	Continue to build coding skills online, then undertake a technology task, such as designing and building a Mars explorer

(continued)

Table 13.2 (continued)

Sch'l	PL#: year level (curriculum content)	Teacher collaboration	In these projects, students…
E	PL1: year 8 science (plant and animal cells)	Science teacher makes connections to STEM practices	Represent the function of Plant and animal cells as two challenges: (1) different objects to represent the different organelles, and (2) bucket of cells
	PL2: year 8 science (particle theory of matter)	Science teacher makes connections to STEM practices	Use representations to (1) physically show temperature effects during chemical changes, and (2) design a communication that shows relationships between reactions, equations, and the particle theory
	PL3: year 8 science (body systems, and movement)	Science teacher makes connections to STEM practices	Use representations to explain the operation of a body system and connections to survival, then present to a younger audience at an expo
J	PL1: year 7 science (simple machines and forces)	Science teachers connect science and engineering	Design a Rube Goldberg machine focussing on forces
	PL2: year 8 science (forces and simple machines)	Science teachers connect science and engineering	Design, construct, evaluate and communicate to others about small design challenges: build a bridge, an make a water rocket
	PL3: year 8 science (forces and energy)	Science teachers connect science and engineering	Design, construct, evaluate and communicate to others about two design challenges: balsa wood bridge, and a water rocket, culminating in a STEM challenge day

Table 13.3 SSSP projects per school in professional learning sequences 1–3 (PL1-3)—integrated teaching teams

Sch'l	PL#: year level (curriculum content)	Teacher collaboration	In these projects, students…
D	PL1 and PL2: year 8 mathematics (generating and using data), science (experimental design)	Maths/science teacher co-ordinates maths and science activities	Design, test, evaluate and re-test a "Barbie bungee" in order to build up problem solving and inquiry skills
	PL3: year 9 mathematics (across a range of topics)	Maths teacher co-ordinates STEM themed extension tasks	Work with a mentor, drawn mainly from tertiary student volunteers, on problem solving tasks
F	PL1, PL2 and PL3 year 7 and 8 science (human senses), mathematics (measurement and variables), technology (robotics and coding)	Science teacher make connections to mathematics and information technology	Develop and apply programming skills to lego EV3 robotics concurrent to doing coding in technology classes. Students compare and contrast human senses and robotic sensors, and link to the electromagnetic spectrum
G	PL1: year 8 mathematics (scale, circles and measurement), science (properties of materials), technology (design and construction)	Co-ordinated approach to design challenge by science, maths and tech teachers	Investigate, design, create and evaluate a vehicle that will travel furthest down a ramp, and represent learning in a portfolio that was assessed in the three subjects
	PL2 and PL3: year 8 mathematics (scale, circles and measurement), science (properties of materials), technology (design and construction), and art and design (graphic design)	Co-ordinated approach to design challenge by science, maths and tech teachers	Investigate, design, create and evaluate a vehicle that will travel furthest down a ramp, and represent learning in a portfolio that was assessed in the three subjects. Includes industry visits where they are informed about modern design and construction practices relevant to the challenge

(continued)

Table 13.3 (continued)

Sch'l	PL#: year level (curriculum content)	Teacher collaboration	In these projects, students…
H	PL1: year 7/8 mathematics (generating and using data), science (experimental design)	Sciences and maths teacher makes connections to science and engineering	Design and construct a number of small projects—parachutes, cranes, bridge building—in order to build up problem solving and inquiry skills
	PL2: year 7/8 mathematics (indices, powers and exponential change), science (acids and bases,	Science and maths teacher makes connections between science and maths	Link three diverse topics from real life (pH levels, radioactivity and the Zika virus). In mini-workshops, students share and communicate as peer tutors
	PL3: year 7/8 mathematics (fractions, decimals and percentage)	Science and maths teacher makes connections between science and maths	Undertake a series of tasks related to real life that gradually build up transferrable cross-curricular skills: paper plane design, construction and testing; transport logistics; and 3-dimensional cube nets. Students then communicate to others through a STEM Expo
I	PL1: year 8 science (forces), technology (design process)	Maths/science teacher makes connections to science and engineering	Design and construct a small self-propelled vehicle to carry water over a set distance, after completing immersion experiences and team building activities
	PL2: year 8 science (investigating), mathematics (basic operations, data and statistics)	Maths/science teacher makes connections between science and maths	Inquire into links between mathematics and science through a series of questions: *How long would it take to watch all the episodes of a number of seasons of a television series?* and design and build a box that has the largest volume from an A3 piece of paper

(continued)

Table 13.3 (continued)

Sch'l	PL#: year level (curriculum content)	Teacher collaboration	In these projects, students...
	PL3: year 8 science (forces), technology (design process)	Maths/science teacher makes connections between science and maths	Design and construct a device that transfers energy (incorporating kinetic and potential energy changes) through a series of changes that eventually rings a bell, then use posters to report findings

Some schools (such as Schools C, H, I and J) are developing a dossier of activities that are embedded into their normal curriculum, while others (such as School F and G) are developing a particular programme and improving it each year. Other schools are developing and extending a pedagogical approach; for School J it is the use of representations (Tytler, Prain, Hubber, & Waldrip, 2013) in Science, and for School B it is open-ended investigations in Mathematics. School D developed a project in PL1-2, but moved towards a system of using Mathematics mentors for their year 9 students thereafter.

By PL4, many of the schools had multiple initiatives occurring at multiple year levels, and included both curricular and extra-curricular activities (School D did not continue into PL4). It is beyond the scope of this chapter to detail the breadth of initiatives emerging, however it is worth noting that over time many (but not all) of the schools were expanding their STEM programmes beyond the projects listed in Tables 13.2 and 13.3.

13.5 Case Studies

In an earlier chapter of this volume [refer to chapter Tytler et al.], we describe the STEM projects of School B. The key learnings of the three Mathematics teachers undertaking SSSP highlight the importance of paying attention to not just the initiatives and activities being produced, but how the conditions within which teachers operate shape teachers' decisions about how to incorporate STEM in a sustainable way. School B is an inner-city girls government school, where the Mathematics background of students entering Year 7 was often weak, and there tended to be low expectations, interest and enrolments in senior STEM subjects, particularly high level Mathematics. In order to respond to this impetus for change, the three teachers developed a common approach to teaching Mathematics that focused on big ideas and core principles using open ended investigations that ran for 2–3 weeks and which linked to students' lifeworlds (Table 13.2 lists the inquiry questions). Student engagement with Mathematics was a key motivator for teachers to take on this activity-oriented approach, and to develop credible and generative assessment processes that reflected the STEM practices of creativity, problem solving, collecting real data and drawing conclusions from evidence.

In order to illustrate the relationship between the emerging projects and the rationale guiding teachers' decision making, STEM projects from three other schools are showcased below: Schools J, G and H. The approaches from these three secondary schools are described to illustrate:

- how the context of each school provides the impetus for change;
- the variety of teaching strategies that schools can use to develop and implement STEM curriculum; and
- various teacher collaboration models that are needed in different circumstances.

Each case study includes background to the school and STEM within the school, the teachers and STEM practices involved, a description of the projects, and how they evolved. In the ensuing discussion, we highlight the importance of considering school context when working with teachers to promote teacher and school change.

13.5.1 Case 1: School J

School J is a young and small private school; however, the present structure, school improvement plan and the majority of staff date from 2014. The curriculum has a traditional structure, with stand-alone, discipline based classes, that is, Science and Mathematics are separate subjects and have no involvement with the Technology faculty. The school works within the Australian Curriculum and pedagogy is predominantly textbook based and exam driven (formal exams every semester). Students often have the same teacher for Mathematics and Science.

The present Head of Science, who was employed in 2015, drove the impetus for change. Reading about, and meeting with, Canadian educators who were active in STEM education and participating in the Deakin University Science and Engineering Challenge, had inspired her. Teachers saw the potential for innovation was being constrained by the lack of a laboratory assistant and a small staff, some of whom were relatively inexperienced. The STEM Programme teachers were the Head of Science/Year 8 Science teacher, a graduate Year 7/8 Science and Mathematics teacher who had made a career change from engineering, and the Year 7 Science/Victorian Certificate of Education (VCE) Chemistry teacher.

The project aimed to change the traditional teaching pattern where theory was fore-grounded followed by illustrative activities, to a model where activities drive student learning and where theory is introduced as needed. The other aim was to introduce Engineering processes into Science, specifically identifying a problem, working in teams to solve the problem, and then testing the solution.

In PL1, the Year 7 Science teachers worked closely to plan and implement a design-based challenge with their Year 7 Science classes where they combined two units of work, Simple Machines and Forces. Then students were presented with the challenge of working in groups to design a *Rube Goldberg machine* to deliberately over-engineer a machine to complete a simple task. The design brief was to incorporate simple machines and identify the forces and chain reactions of energy transfers. The STEM practices (from Table 13.1) addressed were: creativity, problem solving, applying tools to a new context, working in teams and communicating findings.

In PL2, the two Year 8 Science teachers designed and implemented two short, self-contained and hands-on challenge tasks in their Science classes: building a bridge, and designing and measuring the distance of a water rocket. The STEM practices addressed in this sequence were again: creativity, problem solving, applying tools to a new context and working in teams. A preparatory design challenge of building a paper bridge was completed in Year 8 Mathematics classes to prepare the students for problem solving and collaboration, and to develop teachers' scaffolding strategies.

In PL3, the teachers re-developed the design challenges from PL2 so that they were undertaken over a number of weeks and formed a single assessment task. Small groups of students were required to design and construct a water bottle rocket and build and test a balsa wood bridge for strength. The learning sequence culminated in a STEM challenge day in which the devices were tested and communication reports presented and shared. The STEM practices were: creativity, problem solving, communication, working in teams, and collecting real data. A student evaluation showed that teachers' careful selection of student groups enhanced student learning and confidence.

13.5.2 Case 2: School G

School G grew out of a former technical-high school and has successfully rebadged itself with an emphasis on individual student responsibility and a carefully planned teaching and learning environment. STEM subjects were taught as stand-alone disciplines with little cross co-ordination or integration. Participation in the project sought to address both the declining numbers of students enrolling in post compulsory STEM subjects and the tendency for the more capable girls to opt into the life Sciences and avoid the physical Sciences. The SSSP teachers were an inter-disciplinary team of Mathematics, Science and Materials Technology teachers.

In PL1, the teachers were given professional learning time and support by the school. The team decided to adopt a design-based strategy in their projects because it was applicable across the three subjects. The teaching team planned an integrated Year 8 unit that was taught to two classes. The unit sought to expose students to industrial processes and problem solving. Each teacher taught a different component of the unit in their separate subjects, with clear contributions from Science, Mathematics and Technology in solving the common problem: *the Rolling Vehicle Challenge*. The unit also had the objectives of stimulating an understanding of STEM and the career opportunities of STEM related subjects and allowing students to make connections between Mathematics, Science and Technology. An excursion to the local university engineering teaching and research facility was part of the unit and it culminated in a celebration day where the teams from the classes tested and presented their solutions in the presence of guests (internal and external) and competed in a vehicle run-off. Teachers taught in their individual disciplines but there was a conscious cross-curricular co-ordination of learning activities and a common STEM language was used which was made explicit to the students and reinforced by the use of a reflective journal across all three disciplines. Assessment was based on performance of their vehicle in the run-off, and a presentation of their journal that documented their research, their design processes and findings. The project addressed the STEM practices of problem solving, creativity, inquiry, exploring and investigating artefacts, using professional/technical language in communication and working in teams, collecting data and using evidence. Timing and co-ordination of classes remained a challenge and, despite efforts to the contrary, a survey showed that the students often failed to make connections between Mathematics and Science activities.

In PL2, the school began to make connections with a local high technology industry, and widened the scope of the unit to include the Art and Design faculty. The teaching team decided to wait to PL3 to implement the project, and the school team concentrated on setting up links with local companies, re-designing the programme and scaffolding required, and making the links between the project and the curriculum for each of the four curriculum areas involved (Science, Technology, Mathematics, and Art and Design).

In PL3, the vehicle challenge was implemented over 6 weeks and incorporated learning outcomes from the four curriculum areas. Again, the challenge was to design, construct and test a rolling vehicle. The students, within each of the four Year 8 classes involved, were organised into teams of three, with roles distributed as leader, recorder or designer. Expectations included the production of a portfolio covering the development of their vehicle, including testing and evaluation, and a multimedia presentation that summarised the production process. Excursions and incursions were organised to three local companies relevant to modern design and construction practices in the car and cycle manufacturing industries. Exposure to these companies was designed to assist and inform the students in their production of their vehicle. Each excursion had a particular focus: meeting the needs of clients, design and construction, testing, and production processes. A *jigsaw* approach was used where each student from each team attended one of the three companies, and then reported their findings back to their group to inform their vehicle design. At the end of the unit, the best performing vehicles were tested in a celebration day in front of representatives from the companies they visited.

13.5.3 Case 3: School H

School H is a medium sized government secondary school. It runs an integrated, Year 7 and 8 programme, with Science and Mathematics teachers sharing the teaching of combined classes. During participation in SSSP, the teachers were mapping their courses against the new Victorian Curriculum and developing common assessment tasks. The SSSP teachers were the Year 7 and 8 Science and Mathematics team of three teachers. Different projects, mainly as a collection of learning tasks, were developed over the three PL sequences, and have been mapped and documented in keeping with the remainder of the other Science and Mathematics curriculum. In doing so, the work of the SSSP teachers was relevant to the changes already occurring at the school.

In PL1, the school-wide focus on critical thinking formed the basis of the planned learning tasks. The teachers conducted the activities in integrated classes, with the secondary goal of developing stronger links between Science and Mathematics. Emphasis was given to drawing on real life contexts, and the skills required in jobs that related to the hands-on activities. The challenge-based activities were designed to be performed in small groups, and completed in two or three class periods. The challenges, which had a competitive edge, required students to design and construct: a parachute that descends the slowest, a paper crane that lifts 50 g, and the strongest bridge. The STEM practices addressed were: problem solving, creativity, using tools, applying new concepts, working in teams, collecting real data and using evidence to validate solutions. In an evaluation, students indicated greater engagement in both Science and Mathematics, but teachers reported that more effort was required to design suitable assessment rubrics or criteria to match the tasks. As a result of this reflection and evaluation after the unit, for the next year's programme they planned to construct a template or design a common approach to these types of activities so that students could become familiar with the appropriate design processes and problem solving techniques.

In PL2, interesting pedagogies were used where the students tutored their peers in order to promote deeper engagement with the materials and improve the communication skills of the students. Representatives from each group of three or four students participated in teacher-run workshops on three seemingly unrelated Science topics: the Zika virus, radioactivity, and biological growth, but the common thread was the Mathematics topic of indices and powers. In addition to the STEM practices addressed in the PL1 problem solving activities, there was a stronger emphasis of communicating and team work. Each student reported their 'findings' back to their groups. This time, the evaluation indicated that the students had a better understanding of the concepts, had greater ownership of their learning as a result of explaining the concepts to their peers, and were beginning to see links between the subjects.

In PL3, the teachers planned a series of learning activities that were constructively aligned: a mixture of small scale design challenges (such as designing, making and testing a paper plane), Mathematics problems drawn from real life relating to fractions, decimals and percentages (such as relating recipes to a shopping list), and critical thinking exercises (such as problems based on logistics—the efficient movement of people and goods). All activities matched the learning goals for the term, and also became practice exercises for the major common assessment task. The culmination was a STEM Expo, where small groups of students presented posters and other forms of communication to each other. The evaluation was run as a peer feedback opportunity, with pairs of students, one from Year 7 and the other from Year 8, visiting each 'booth' at the Expo and providing peer feedback via a teacher designed reflection sheet. The same STEM practices were addressed as for PL2, as well as the application of ideas to new contexts that was promoted through the STEM expo as students gave feedback to each other.

13.5.4 Discussion

Evident in the three cases, and School B, are commonalities and differences in the teaching strategies used and how the STEM subjects were included. All schools incorporated problem solving in one form or another. Other teaching strategies used included, for example, open ended investigations (School B), design challenges (Schools J, G and H), peer teaching (School H), jigsaw approach (Schools H and G), and group work (Schools J, G, H). Schools H and J developed different activities for each PL sequence, and demonstrated increasing sophistication in the complexity of the tasks and links between the subjects. School G also found that students were more successfully able to make links between the subjects, especially in the second iteration of their vehicle challenge, partly because the teachers were more aware of the need to make the links explicit for students.

Different teacher collaboration models are also illustrated across the four schools, with a Mathematics team (School B), a Mathematics-Science team who integrated Mathematics and Science (School H), a Mathematics-Science team who taught both subjects but separately (School J), and an interdisciplinary team of teachers who taught one of the subjects (School G). This variety illustrates that there are many ways to incorporate STEM into schools: both discipline-bound and integrated approaches to STEM can lead to effective learning outcomes; STEM curriculum and the STEM practices can be taught by a single teacher or a team of teachers; or STEM curriculum can be a substantial learning sequence based around inquiry, or contained activities.

This variety in responses to the STEM 'push' is measured by the needs of the schools, and they are reflected in the reasons the schools became involved in SSSP. All schools indicated an intention to focus on improving student engagement. At School B, for example, the intention was to respond to low engagement of the girls in Mathematics, especially at the senior level, therefore, enhancement of the Mathematics curriculum was the intended focus. At Schools G and H, a desire to improve students' ability to make links between different subjects was the driver to integrate the subjects. Also important at both schools was to bring 'real life' into the classroom, where, at School H this manifested as real life problems recognisable for students, and at School G, links with local companies brought the world of work into the classroom. At School J, the potential for STEM to inspire through student directed project work was regarded as a way to engage students; also, the teacher support provided through SSSP was a motivator to participate in SSSP as it was seen to attend to the lack of resources and teaching experience at the school.

The 2016 report[1] by an independent evaluator highlighted a number of factors that have been critical to the success of SSSP thus far. Three of these factors are outlined below.

13.5.4.1 Teacher Time Release

Teachers found that the most valuable aspect of the project is funding for teachers to have time for guided planning, both at school and during the professional learning days where there is cross-fertilization of ideas across schools. Time is needed for teachers to plan for and implement STEM learning, generate and analyse evaluative data, reflect on their learning, then collaborate with teachers to re-develop and develop new programmes. Time is needed to up-skill and recruit new teachers to these new directions and pedagogies. Time is also needed to embed, sustainably, a STEM focus into the curriculum, strategic plans and future directions of the school. For students, time is needed for them to learn, hone their skills, and re-consider their career aspirations; longitudinal data is required to capture changes in student subject choice and destination data.

13.5.4.2 STEM Expert Support

SSSP, particularly through the Project Officer, provides invaluable STEM expert support for teachers in the schools. As a STEM enthusiast, the Project Officer worked closely with many of the teachers to develop new curriculum, supported them in running some activities, and even prepared equipment. The schools used the Project Officer in this capacity to different extents, some often, others rarely, depending on the needs of the teachers and time they were devoting to developing new curriculum. The availability of the Project Officer was noted in the evaluator's report as being a key success factor. Support through the professional learning days was also noted by the evaluator as building teacher confidence and capacity to design and implement effective STEM programmes, and develop directions for the future of STEM in their school. In a number of cases, the sharing of projects resulted in schools adopting and adapting the ideas emerging from other schools, for example, the STEM Expo at School H had been inspired by a STEM Expo at School E involving children from the feeder primary schools.

Further, a number of teachers have taken on positions of responsibility in relation to STEM since starting the programme, and a number reported that they are being called on by school leadership to help shape the future of STEM in their school.

[1] Teacher and program leader interview, meeting outcomes and project outputs were used by the evaluator as evidence.

13.5.4.3 STEM Academic Leadership

Deakin's involvement has been crucial, both in providing the supportive personnel and infrastructure, but also in setting a programme focussing on and documenting teacher and school change. STEM academic leadership is important for a number of reasons.

Academic leadership enables a research informed and co-ordinated approach to providing a common language around STEM and flexibility in responding to and meeting the respective needs of each school. It became evident during the reporting days that teachers were adopting the common language relating to the STEM practices and the different STEM pedagogies. The variety in how schools approached STEM and the different processes involved in reflecting on and re-developing or diversifying their projects illustrates the need for flexibility in professional learning for STEM teachers. While many other STEM initiatives focus on providing packaged STEM learning experiences for students, or raising teachers' awareness of and access to STEM activities and programmes, the SSSP Programme takes a flexible, school-directed approach to building an expectation of sustained, school-generated, evidence-based change. The evaluator noted that this flexibility remains a key strength of the programme, both in structure and execution; the negotiation between the teachers and the academic leads and Project Officer have been critical to this flexibility and can only occur where there are ongoing and trusting relationships that are built over time.

Academic leadership can be provided to schools for generating a body of evidence of the impact of the school-based initiatives. Documenting such evidence, as promoted by the *National STEM Education Strategy* (Education Council, 2015), can, for example: mobilise future buy-in from other teachers at the school, the school parent communities and other schools; attract the attention of governments/policy makers and funding bodies; and provide evidence to support career advancement for the participating teachers.

13.5.4.4 Challenges

One of the challenges that schools are still wrestling with is developing appropriate assessment practices. Schools have used a range of activities for assessment, such as student journals/portfolios, presentations, artefacts, and worksheets. However, the method and focus of assessment are still being developed by some schools, with tensions arising between: criterion based assessment versus descriptive assessment; a focus on disciplinary content versus STEM practices or inquiry processes; and a focus on reporting versus a focus on student engagement. Decisions made in response to these tensions have depended on whether the STEM programme is part of the mainstream subjects (such as part of Science, Technology or Mathematics units) or integrated across a number of units, or whether they are stand-alone STEM experiences. For example, School B uses learning intentions focussing on students' ability to apply Mathematics concepts to real world problems, rather than explicitly assessing mathematical skills and conceptual understanding.

There are challenges, or 'common barriers' (Hackling, Murcia, West, & Anderson, 2014, p. 9), associated with school-industry collaborations when attempting to make links between school content and the world of work. School G's partnerships with three local industries are a successful example of how schools can embed meaningful learning experiences in a way that connects the school project of building a vehicle with three relevant local industries—the local car testing facility, a high-end cycling manufacturer, and an aerospace manufacturer. One of the challenges that can arise, however, is that teachers and industry representatives often do not share a common language, and the language of industry may not be understandable for students. This can be alleviated when a teacher has some knowledge of the industry and can be selective in which industry practices are emphasised and inserted into the learning tasks, or if a 'broker' is engaged who understands both languages and can help to translate each partner's needs and offerings. Other issues that can arise relate to the time needed for partners to undertake these negotiations, as well as carry out the activities such as through excursions or incursions. Hackling, Murcia, West, and Anderson (2014) found that submissions to their inquiry identified the following as potentially addressing barriers to establishing industry-school partnerships: 'better co-ordination, administrative support, and working in partnership with education service providers who understand how schools work and have specialist curriculum knowledge' (p. 9).

There are long-standing challenges around maintaining the integrity of the disciplines during curriculum integration (Rennie, Wallace, & Venville, 2012). There can be a tendency for one or two of the disciplines to be fore-fronted with Mathematics, and sometimes the Science, being used as tools instead of the disciplinary ideas being explicitly taught. Conceptual development can be compromised and sidelined in preference for other outcomes such as engagement, with creativity, design processes and construction being the focus of assessment. Key to maintaining the intellectual rigour of the STEM challenges and activities is being explicit about how the STEM practices and Science and Mathematics concepts are integrated into the unit or learning task. Mapping of curriculum (e.g., as was done by Schools H and G) can be important in explicating how the disciplines are realistically represented.

Another challenge is convincing school leadership, other teachers, students and even parents of the value of rethinking the curriculum to include STEM learning opportunities. To do this, evidence of change to student learning outcomes and student engagement can be fundamental to the ongoing acceptance and thus successful embeddedness and sustainability of STEM learning experiences in a school more accustomed to a traditional siloed approached to curriculum. Apart from the data gathered by the Deakin team, teachers are beginning to embed opportunities to gather data. While they have collected data from students during each sequence, they are developing more sophisticated tools to find out the effects of the programmes on student learning, engagement, and other variables considered important by the schools. Further support for teachers researching their practice is the focus of PL4.

13.6 Conclusion

State and federal education authorities have reacted to the STEM 'push' by initiating a range of policy changes culminating in the *National STEM School Education Strategy* (Education Council, 2015) that aims to: raise student STEM participation and achievement through increasing student aspirations; improve teacher capacity and quality; support within school systems; create partnerships with tertiary providers, business and industry; and build an evidence base. These aims are comprehensive and resonate with initiatives in other parts of the world, such as the European Community where attempts have been made to raise student STEM awareness, establish industry and school links, and build up STEM teaching skills (Scientix, 2014).

SSSP, too, meets these aims by supporting teacher learning through targeted intensive and ongoing support to develop inspiring and interesting STEM learning experiences, where traditional silo bound and textbook teaching approaches give way to problems from 'real life' and serious engagement with the world of work. Critical to the success of this programme was that it was longitudinal in nature with a combination of formal professional development with ongoing support, fostered was a community of teacher learners who were willing to share and take risks, the programme was designed from the outset to respond to teacher and school needs rather than as a 'one-size-fits-all' approach, and teachers were given time and space to reflect, plan and change beliefs and practice. Where previous large or medium scale interventions use professional development based on delivering specific programmes or practices, unique to SSSP has been the encouragement of schools to develop their own approach to conceptualising and implementing STEM that makes the best use of their resources. The outcome has been a wide variety of projects organised in schools in different ways. What holds the programme together has been a comprehensive discourse of STEM through the STEM Vision framework modelled during the PL intensives and by the programme team, and which over time helped the teachers to develop rich practices in their own particular ways and contexts.

The first step for any teacher or school in embracing STEM is to articulate what STEM is and needs to be for their particular context; the seemingly amorphous nature of STEM is both its power in being shaped to potentially provide a solution to some of the problems faced by schools (such as student disengagement), but also makes it difficult for teachers to navigate without guidance and support, hence the demand for professional development in this space. A common but flexible language around practices and pedagogies suitable for STEM education must be articulated and made explicit in this professional development if sustained teacher and school change is to be achieved.

References

Clarke, D. J. (2015). Comparative research in mathematics education: Boundary crossing and boundary creation. In K. Beswick, T. Muir, & J. Wells (Eds.), *Proceedings of the 39th Conference of the International Group for the Psychology of Mathematics Education*, vol. 2 (pp. 169–176). Hobart, Australia: PME.

Education Council. (2015). *National STEM school education strategy: A comprehensive plan for science, technology, engineering and mathematics education in Australia*. Canberra: Education Council.

English, L. D., & King, D. T. (2015). STEM learning through engineering design: Fourth-grade students' investigations in aerospace. *International Journal of STEM Education, 2*(14), 1–18.

Hackling, M., Murcia, K., West, J., & Anderson, K. (2014). Optimising STEM education in WA schools. Part 1. Summary report. Technology and Industry Advisory Council. Joondalup: Edith Cowan Institute for Education Research. http://www.tiac.wa.gov.au/Files/STEM_Report_Part-1_20022014.aspx.

Hobbs, L., Cripps, Clark J., & Plant, B. (2018). Successful students—STEM program: Teacher learning through a multifaceted vision for STEM education. In R. Jorgensen & K. Larkin (Eds.), *STEM education in the junior secondary: The state of play* (pp. 133–168). Singapore: Springer.

Rennie, L., Wallace, J., & Venville, G. (2012). Exploring curriculum integration: Why integrate? In L. J. Rennie, G. Venville, & J. Wallace (Eds.), *Integrating science, technology, engineering and mathematics: Issues, reflections and ways forward* (pp. 1–11). New York: Taylor and Francis.

Scientix. (2014). *The community for science education in Europe*. Brussels: European Schoolnet.

Tytler, R., Prain, V., Hubber, P., & Waldrip, B. (Eds.). (2013). *Constructing representations to learn in science*. Rotterdam: Sense.

Chapter 14
"Draw What You See" Transcending the Mathematics Classroom

Signe E. Kastberg, Rachel Long, Kathleen Lynch-Davis and Beatriz S. D'Ambrosio

Abstract To gain insight into ways student's experiences with mathematics can support them to reach their human potentials, we explored children's engagement in a collaborative art project. We describe the teacher-developed project and facilitation approaches that supported the exploration. Using a narrative inquiry methodology and artefacts from the experience, we narrate children's experiences using four dimensions: autonomy, authority, success, and relationships with others. We contend that the children were involved with ideas and peers in ways that resulted in building a positive relationship with mathematics, producing a counter-narrative to one of failure and helplessness typical of mathematics as a discipline. Recommendations for further study focussed on mainly mono-disciplinary contexts are shared.

Keywords Mathematics · Art · Student collaboration · Affective behaviours

14.1 Introduction

In today's school climate, the teaching of mathematics has acquired a sense of urgency, which has deterred teachers from creating joyful opportunities for engaging in mathematics. Rarely are students encouraged to be curious, posing and investigating their own questions. Such activity is often seen as drawing away from the more

Beatriz S. D'Ambrosio: Deceased October 21, 2015.

S. E. Kastberg (✉)
Purdue University, West Lafayette, USA
e-mail: skastber@purdue.edu

R. Long
Central Elementary School, Beech Grove, USA

K. Lynch-Davis
Texas A & M University Corpus Christi, Corpus Christi, TX, USA

B. S. D'Ambrosio
Miami University, Oxford, OH, USA

B. Doig et al. (eds.), *Interdisciplinary Mathematics Education*,
ICME-13 Monographs, https://doi.org/10.1007/978-3-030-11066-6_14

urgent task driving instruction, which is to learn the mathematics that will result in gains on achievement tests. D'Ambrosio and D'Ambrosio (2013) assert that the primary goal of education is to support individuals to reach their human potential. The discipline of mathematics serves this goal, as it takes on many forms in human activity beyond the completing of exercises with a single answer often seen in mathematics classrooms and on achievement test. Interdisciplinary mathematics education (IdME) (Williams et al., 2016) provides a lens through which to view the complexity of mathematics in human activity.

In this chapter, we discuss the landscape of investigation, as described by Skovsmose (2001), and opportunities for exploring mathematical ideas and building human relationships provided in such contexts. We take these explorations outside the mathematics classroom and examine a landscape of investigation drawn from the art classroom. Using narrative inquiry (Clandinin & Connelly, 2000; Polkinghorne, 1995), we characterize fourth-graders' (ages 10–11) experiences in the "mainly mono-disciplinary" context (Williams et al., 2016, p. 19) of the art class using four dimensions: authority, autonomy, success, and relationship with others. The democratic environment created in the art classroom supported children to find and use their mathematical voices, as they identified and overcame social and mathematical challenges by negotiating and co-constructing ways of knowing mathematics to complete the project.

We contend that the children were involved with ideas and each other in ways that contributed to building a more positive relationship with mathematics, producing a counter-narrative to one of failure and helplessness typical of mathematics as a discipline. We conclude by discussing opportunities and challenges involved in increasing the level of complexity of this landscape of investigation "on the continuum of relations between disciplines" (Williams et al., 2016), while maintaining support for children to build mathematics identities and power in their efforts to reach their human potential.

14.2 Literature

Those who use the rhetoric of "mathematics for all" and defend the idea as a goal of mathematics education, assert that all students should have equal opportunities and resources to learn rigorous mathematics. For D'Ambrosio (1990) the teaching of mathematics occurs within the structures of schooling and education, and should contribute to achieving the lofty goals of education where:

> Mathematics education ought to prepare citizens so that they will not be manipulated and cheated by indices, so that they will be allowed to change and to accept jobs which fulfil and appeal to their personal creativity; that is, so that individuals will be allowed the satisfaction of their own creativity and will be free to pursue personal and social fulfilment thus being able to achieve happiness. (p. 21)

Learning mathematics in this way, results in supporting individuals to achieve their human potential. Contrary to this goal, school mathematics:

> fulfils social functions of differentiation and exclusion… Instead of opening opportunities for all, mathematics education generates selection, exclusion, and segregation. A demarcation is established between those who have access to the power and prestige given by mathematics and those who do not. (Skovsmose & Valero, 2001, p. 41)

As students experience school, they come to embrace their success as young mathematicians, or accept that mathematics is beyond their comprehension. This troubling reality oppresses learners into limiting their dreams and aspirations from a young age. It is beyond the scope of this paper to explore the reasons that students fail at school mathematics. Many scholars have dedicated their careers to understanding this situation and proposing alternative educational experiences. For the purpose of this paper, we have chosen to highlight the work of scholars who propose to enable all students as viable mathematicians. Central in this proposition is providing experiences that will empower the young to be citizens who engage in finding creative solutions to the social and economic injustices of society, by understanding the world in a critical and conscious way (Gutstein & Peterson, 2005). Real-life references for meaning making "seem necessary in order to establish a detailed reflection on the way mathematics may be operating as part of our society" (Skovsmose, 2001, p. 131).

A democratic mathematics education process that allows all students to fulfil their human potential would include opportunities for students to develop mathematical voices as they describe what they know and how they understand the world. Students' experiences in the discipline of mathematics shape and define their identities as mathematical beings (Walshaw, 2005). In addition, their perceptions of how others view them, plays an important rôle in shaping their identity. It is in a climate of inordinate amounts of testing that young people's identities are currently taking shape. As such, they perceive their "value" as learners by their scores on standardized tests. These quantitative views of their value leave them feeling dehumanized and lead to negative relationships with the subject matter.

According to Freire (1998), teaching begins with respect for what students know. When their existing knowledge is the foundation for learning, students will acquire autonomy and agency. Considering democracy as social interactions of individuals relating to each other as they collaborate to "produce their cultural and material living conditions" (Skovsmose & Valero, 2001, p. 46), allows us to interpret classroom settings as microcosms of society where such interactions occur. In such environments, students learn to speak up, question others, and defend their thoughts, while maintaining a relationship of kindness, care, solidarity, and respect. In the art class we describe, the students collaborated to create a cultural artefact and negotiated decisions with respect for each other's ideas while attempting to create mutual understanding. We interpret this experience as supporting the development of a key component of children's participation in a just and equitable society. This is the view Signe and Rachel held while in the classroom with the children, and that the research team later used to analyse field texts and construct a narrative.

We tell the story and analysis of our experience through narrative.

> Narrative inquiry, the study of experience as story, then, is first and foremost a way of thinking about experience … Narrative inquiry as a methodology entails a view of the phenomenon. To

use narrative inquiry methodology is to adopt a particular view of experience as phenomenon under study. (as cited in Clandinin, 2006, p. 45)

Narrative inquiry is a methodology "for studying lived experiences" (Clandinin, 2006, p. 44), resulting in stories that illustrate how people experience and make meaning of the world. In our situation, Signe entered in the midst of a story, and lived the experience alongside Rachel, the art teacher, and the children. Signe and Rachel created field texts by making a digital photo record of the experience. They then added to the field texts by writing about the experience and by discussing their observations of the children's work and interactions. During weekly meetings the research team discussed the continuing events in the art classroom to gain insight into the children's actions and how Rachel and Signe were supporting these actions. Our collection of "descriptions of events and happenings" (Polkinghorne, 1995, p. 12) served as the data and our conversations provided early insights into the critical actions taken by Rachel and the children in the production of the cultural artefact. Analysis of the field texts and images was recursive as we developed "a plot that serves to configure data elements into a coherent story" (Polkinghorne, 1995, p. 16). We read and re-read field texts and images, building from initial views of important actions, taken by Rachel and the children, toward a narrative explanation of what happened that contributed to the collaborative construction of the cultural artefact. In addition, through weekly discussions, we negotiated, collectively, the way in which we would narrate the story of the children's experience. This methodology also allowed us to analyse the field texts once away from the classroom. These conversations involved looking for "connections of cause and influence among the events" (Polkinghorne, 1995, p. 18) and the reasons actions were undertaken by the children. Through this process, we identified four dimensions that we use to narrate the children's experiences from our perspectives as researchers. The viewing and reviewing of the events, the many questions we asked of Rachel and Signe, the many questions they asked of themselves, and the return to the classroom to ask the children questions as well, resulted in our deeper understanding of the experience. Our narrative contains digital photos and excerpts from field texts that help represent dimensions of the children's experiences, as we understood it. The narrative we created reflects the shifting landscape of our understanding of the children's experience, as we lived, interpreted, and continue to make it meaningful personally. Importantly, the narrative reflects our understandings of the children's experience, not the childrens' own understandings of their experience.

Typically, educators in the discipline of mathematics focus on the use of numbers in the fabric of life, overlooking other mathematical content that is part of that fabric. Yet in landscapes of investigation, contexts serve as invitations to students "to formulate questions and to look for explanations" (Skovsmose, 2001, p. 125). In the work that we describe from an art class, children seem unaware of the mathematical qualities of their actions. They seem liberated from their imposed (quantitatively informed) mathematical identities and anxious to engage in a creative act, so different from computational work typical of mathematics classes. Yet listening to discussions in the art class revealed language of space and lines, and the management of both, as the children overcame obstacles involved in scaling. Such challenges and related reasoning have been explored in mathematics education (for example, Streefland, 1984; Cox, 2013) without explicit attention to development of notions of identity and power, possible, in such contexts. In the art class, we describe the mathematics involved as embedded in actions and conversation that also supported the development of identity and power. Teaching in this context supported children in ways that moved them toward their human potential, constructing opportunities for them to create with and enjoy using mathematical reasoning, to engage in conversations that involved genuine mathematical talk, and to appreciate the beauty of mathematical patterns and relationships.

We anticipate that in listening to stories of the experiences of the children, we will begin to imagine different possibilities for stories in mathematics classrooms involving dimensions of the discipline of mathematics that include, but extend beyond, knowledge and knowing as the computation of a single answer. We believe that narratives of the experiences of the children will allow mathematics educators to build insights about children's learning of the discipline of mathematics outside mathematics classrooms that attend to children's learning of mathematics embedded in action (Skovsmose, 2001). These narratives may have the power to encourage children to re-story their experiences with mathematics as more positive, more engaging, and more relevant to their lives.

14.3 Context and Task

In art, viewed as a mono-discipline, inspiration and creative voice are expected and anticipated. These inner voices give rise to opportunities for the development of technical skills as well as emotionally provocative creations. In her planning and interactions with the children, Rachel searched for ways to bring different materials to life so the children could have opportunities to explore and discover with different mediums. She wanted the children to feel accomplished and proud of their artwork. She assumed that the critical thinking skills and decision-making the children do in art is carried with them into other academic areas in particular the discipline of mathematics. For Rachel and the research team the disciplines of art and mathematics involved problem solving, meaning seeing different possibilities and answers to dilemmas, yet methods of inquiry in these disciplines vary tremendously.

In 2013, Rachel sought to "create new ways of incorporating math[ematics] into art projects without the children feeling like they are actually doing math[ematics]." She wanted to "find a bigger project for the children than just using a 9 by 12 inch paper" and "have the children do their own critical thinking." This effort was motivated by Rachel's desire to contribute to the children's mathematical development. Rachel found a lesson that involved scaling up a version of Warhol's Campbell's Soup Can (Stephens, 2013). The children worked in small groups of three or four, during one 45-minute period per week for 8 weeks. Each group was given one 8 ½ by 11 inch [approx. 21.5 cm by 28 cm] picture of a "sectioned" can (see Fig. 14.1). The children

Fig. 14.1 Sectioned soup can

were asked to fill in the can with complementary colours and then to cut the can into sections. Each section was then to be recreated on an 8 ½ by 11 [21.5 by 28] sheet of paper. After reassembling all nine scaled up pieces, the children were to colour the artwork and present the work to their classmates for discussion. The final goal was to use all the enlarged cans to create a mural for the school by placing the cans alongside one another. Each enlargement resulted in a version of a can that was roughly 2 feet wide by 3 feet [60 cm by 90 cm] tall and a mural that was 10 feet [300 cm] wide.

14.4 Narrative of Experience

In this section, we highlight the four dimensions that helped us describe and make sense of the children's experience. In particular, the concepts of authority, autonomy, success, and relationships with others take on new meaning in the context of the art classroom. For us, the collaborative experience in which children were co-constructing their art product and making sense of the underlying mathematics involved, was unique. Throughout the experience, the children set parameters regarding when they had succeeded in completing the task, when they had a product they could be proud of, and what explanations would reveal their collective thinking and decision-making. In the environment of the art classroom, the children authored "narratives of confidence and capability, seeing themselves as powerful and whole" (Paley, 2011, p. 745). These narratives are in stark contrast to the ones many live in the mathematics classroom. They are counter-narratives to ones of failure and frustration, inadequate achievement, and disenfranchisement.

14.4.1 Redefining Authority

In mathematics, the "world-view and the knowledge system of the educator" (Freire, D'Ambrosio, & Mendonça, 1997, p. 8) is often used to manage the space that develops for learners to explore, choose, decide, and explain, thereby constraining children's opportunity to use their intuitive sense, to choose a path, and learn from errors. In the soup can project, while Rachel's initial description of the project framed the inquiry space, the children used their reasoning and communication skills to engage in exploration and explanation of the landscape of investigation Rachel described. Their actions and discussions became a real-life reference for mathematics learning, what Skovsmove described as a learning *milieu* (2001). The landscape of investigation, initially described by Rachel, quickly became a landscape redefined by the children. The children, rather than Rachel, were authorities on their work as they chose paths for the investigation and managed social interactions.

Negotiations among group members started before they began scaling up the picture of the soup can, as groups worked out different sensitivities and preferences of group members. Choosing colours and creating a way to colour together on one piece of paper was a problem solved in a variety of ways. One group debated the different colours they might use and then, initially, took turns with crayons. This was not satisfying since all the group members wanted to colour. To give everyone a chance, one child found crayons of similar colours and passed them out. With hands on the same paper, each child began colouring a section of the picture. Heads and hands often touching, the children talked about life, the result of their colouring, and their sense of satisfaction with their work.

In the beginning the children split up the task with each child taking responsibility for enlarging a section of the picture. As a child finished one section, he or she moved on to another. When all sections were complete, problems emerged. With pieces laid side-by-side, different lines and spaces of the sections were not aligned (see Fig. 14.2). Disappointment, gave way to brain-storming how to move ahead. With the class ending and not reconvening for another week, the children put away their work and put aside the debate. The interruption gave the children time to build a strategy for solving the alignment problems that continued to emerge as one artist made adjustments to a section of the picture without consideration, yet, for the work of other artists. After a few weeks several artists began to co-ordinate activities. Signe sat watching one group of boys for the greater part of a class period, documenting steps and mis-steps that they identified. They answered her questions about their actions and decisions patiently. But when they felt she had stayed too long, one boy suggested: "You should go watch someone else now. You can talk to us again next class." He was an authority on what the group needed to do in order to move the work forward. Signe left following Rachel's practice of moving about and just watching or asking questions about decisions. As promised, during the next class the boys again invited Signe to explore their work and patiently answered all her questions about decisions they had made in her absence. The children had the authority to create

Fig. 14.2 Aligning sections

their own paths and make decisions when challenges arose, or when prying eyes and words became an impediment to their work.

In each phase of the project the children made choices and accepted the challenge of, and the responsibility for, decisions they had made. This approach to authority was based on Rachel's respect for the children's process. She understood the need to attend to children's learning and awareness of how they moved from common sense to creating ideas "uncommon" to adults (Paley, 1986, p. 127). The reality of the children and the remaking of a model of that reality through the soup can project involved a redefinition of authority that led to a growth of intellectual autonomy in the art class.

14.4.2 Redefining Autonomy

Students in schools often rely on the teacher to define how they should think and behave in the mathematics classroom, diminishing their autonomy. By redefining authority, so that children have a chance to manage social interactions of the group, choose a path, and learn from errors, teachers can promote and sustain intellectual autonomy among their students. In classrooms where teachers attend to the landscape of investigation and provide authentic ways for students to engage in mathematics, intellectual autonomy develops. They "are aware of, and draw on, their own intellectual capabilities when making mathematical decisions and judgments as they participate in these practices" (Yackel & Cobb, 1996, p. 473). Intellectually autonomous children, engage in judgments and negotiation of what is right and wrong in the *mathematics*. The teacher facilitates students' development of intellectual autonomy by creating environments where inquiry is valued and mathematical discussions amongst students are paramount. Intellectual autonomy differs from children's authority in that the focus is mathematics and mathematical decision-making.

Many instances of intellectual autonomy emerged in the completion of the art project. Rachel, in developing and attending to the landscape of the investigation, nurtured intellectual autonomy by encouraging the children to design their work plan and decide when their scaled picture of the can was acceptable. The children did not ask Rachel or Signe about scaling or how to scale. They called an adult when tools were needed or to share a new insight or achievement. Even when Rachel encouraged one group of children, struggling to align enlargements of the three sections in the first row of the picture, to "draw what you see, not what you know," the children interpreted this advice in ways that made sense for their intellectual work. Rachel and Signe saw the children's problem as one of "seeing" the lines and spacing, yet the children reasoned that their initial approach to scaling, each child scaling a section or two, had created the problem. To solve the problem, they co-ordinated their efforts. They first enlarged adjacent sections (two sections at a time) and then reasoning that coordinating three sections containing a continuous line or lines (such as those on the top of the can) was necessary. While Rachel's suggestion provided a welcome break from the challenge of aligning the enlargements, the children used their intellectual autonomy to revise their approach to scaling rather than Rachel's suggestion.

Intellectual autonomy was evident throughout the project, individually and collectively. When children employed their own capabilities and judgment to draw their portion of the scaled up version of the can picture, they exercised individual intellectual autonomy. Their negotiations within the group as they brought "together the pieces" and resolved discrepancies within the scaled up image that arose from piecing together individual work, illustrated their collective intellectual autonomy.

14.4.3 Redefining Success

Student's participation in school often results in feelings of dis-empowerment as they feel unable to live up to the standards established by the teacher, schools, or districts. Continually falling short, according to norms that assume that all students will learn a list of topics at the same pace, in the same order, and by a certain date in the school year, leave students feeling helpless. Students come to believe that the goal of the discipline of mathematics is to learn procedures, rather than to learn when to use mathematics to develop and solve problems. They rarely, if ever, experience the exhilarating feeling of identifying and defining problem situations and solving problems by using creative approaches to challenging dilemmas.

In the art classroom, as the children worked, there were many opportunities to discuss their work, negotiate what was correct, and consider what needed to be fixed. Children seemed to use visual cues and their intuitive sense of aesthetics to judge their success and decide on directions for revision. The children relied on themselves and on each other to decide whether to continue along a path or to alter the direction taken in their work. There were many instances where Rachel and Signe turned to each other and said, "we will need to wrap this up," since the children seemed frustrated and no longer interested in the inquiry or dilemmas they were trying to resolve. Yet each week, the children returned to the investigation with new energy and ideas, regarding how to solve a problem from the previous session.

Together the children created, although not explicitly, criteria for success. With each decision to alter or fix their product they approached something that they considered better and more pleasing. Their satisfaction with their work was evident in the whole class discussion at the conclusion of the project. Each group of artists had an opportunity to describe their work and the decisions they made along the way.

The children spoke with conviction when they were asked about decisions they had made or characteristics of the can that others had not included. For example, one group had a black spot on the lid of their can (see Fig. 14.3). When asked about the meaning of the spot, they quickly noted that it was an opening. The children reasoned that cans are meant to be opened and when they are, there is often a hole in the top. This hole illustrated the authenticity of the picture in their enlargement. In addition, the children shared what they understood and what still confused them. They were sure that you could align sections of the can by lining up not two but three pieces in a column and checking for alignment. Lining up two sections at a time had caused problems for all the groups and they were quick to point to this challenge, ways they had overcome the challenge, and to point out evidence of their success on their art work.

Children constructed the definition of success implicitly, and used it to describe their work. Use of their criteria to describe the work, was evident in their pride and justifications. Peers' appreciation for each other's work, effort, and specific details of the art, was authentic and full of respect and admiration. Children expressed their delight in the opportunity to hear from others about challenges they had experienced in enlarging the picture and how they had overcome them. For example, one child asked another group who had aligned the curved lines on the top of the can, "How did you get the lines on the lid of the can to match up so well?" The group explained that the lines were drawn and checked and erased "again and again until we were done." Their response was delivered with a tone of satisfaction and humility in their acknowledged success, characteristic of this final discussion.

Fig. 14.3 A group's enlarged can

14.4.4 Redefining Relationships with Others

Rachel posed the project to the children as a collaborative effort. After initially splitting the scaling task, the children began to discuss problems and collaborate. Throughout the interactions, there was evidence of caring relationships among the children, as they listened to each other in kind, respectful ways. The children worked productively together, building understanding and voicing their observations, evaluations of the progress they were making, and proposing ways to improve the product. While the children did not formally spend time establishing norms for working together, they behaved in ways that adhered to implicit norms of respect and care for each other's ideas. They did not agree on all dimensions of their work or decisions made, but they discussed possibilities, tried things out, and came to an agreement of the direction that seemed most reasonable.

Some of the most difficult moments for groups occurred as they put the pieces together and realized there were problems with alignment of the sides of the can, or the top of the can, or—what proved most difficult to all—with the lettering. Collectively, the children made decisions of whether to erase and redo the enlargements, or come up with another strategy for fixing the problem. These were often tense discussions beginning with conflicting claims about whose enlargement was best. Yet, the children moved to resolve differences by removing themselves from the discussion for a time or negotiating which enlargement could be called "best," ultimately looking for evidence of the relative positions of the lines on each section to decide and overcome their differences of opinion.

The lettering challenged all the groups. During final presentations, when Signe asked the children to identify the most difficult parts of the project, the children pointed to the lettering. They also accepted not quite knowing "how to do that [cursive writing] yet" and complimented those whose work showed some measure of success. For example, peers considered the lettering of "SOUP" on the can in Fig. 14.3, as "really good." The presenters explained how they had accomplished this success by having a single artist work with the three sections that contained the letters, working exclusively on the lettering. They acknowledged the expertise of the group member by giving this one artist the responsibility—and the honour—of doing the lettering himself. Individual group members were recognized for artistic talent or extra patience for the task of reproducing and enlarging the lettering of the original artist.

Throughout the project, each child assumed rôles that aligned with their personal talents. Peers recognized ways in which specific group members could contribute to producing a more appealing and successful product. Group members shared power equally and leadership shifted with each new challenge or action. Both comradery and solidarity were evident in the group interactions.

14.5 Discussion and Conclusion

In the art classroom children worked collaboratively to enlarge Warhol's painting of a Campbell's Soup Can. In the process, they refined their perceptual understanding of scaling and proportional reasoning. While the primary goal was to create a mural for the school, intermediate goals developed and spurred side inquiry. Of particular interest to IdME are opportunities within this landscape of investigation that could be further explored within the discipline to increase the level of complexity of the enquiry. In the art investigation, the children's challenges with the lettering on the can, is one such opportunity. The letters are examples of what has been described as internal marks called "secondary lengths" (Cox, 2013, p. 8). Such lengths, in proportional reasoning, must be situated in relation to "primary lengths" such as the height and widths of the can and "gap lengths," implicit lines between primary and secondary lines. Cox describes intuition involving these lengths as standing at the intersection of geometry and number in the discipline of mathematics. In scaling tasks such as this one, according to Cox, students focus on primary lengths and work to relate these to secondary lengths, thereby developing insight into gap lengths. In the art class children were successful in developing strategies for managing alignment issues with exterior lines (primary lengths), but were still dissatisfied with their approaches to internal letters (secondary lengths).

Dissatisfaction with the lettering involved in the investigation opened potential opportunities for movement beyond the use of intuition toward more formal approaches to scaling the lettering such as measurement and computation. Although a child at one point during the investigation, began to use a ruler to measure primary lengths from sections of the enlarged picture, when he was queried about his activity by his group mates, he could not yet describe why measuring might help. This opportunity might have served as a jumping off point for explorations to answer the question regarding the utility of measurement. Skovsmose (2001) identified such opportunities for consolidation of more formal mathematics and mathematical ways of knowing in action during the project as creating "harmony" (p. 129) between modes of activity within the discipline. Further, Guitérrez (2013) suggests that considering the use of tools, such as measurement in formatting realities in projects like the soup can, would provide children with opportunities to think "critically about the benefits and drawbacks" of using mathematics, thereby developing insight into how and when they would want to deliberately use/create mathematics in their everyday lives (p. 47).

While adding a mathematics disciplinary exploration of measurement as it related to the challenge of alignment in lettering could have increased the disciplinary complexity (Williams et al., 2016), what is unclear are the ways in which such an addition might have had an impact on the dimensions of experience of the children. The central goals of the art inquiry involved movement toward the construction of the enlargement. Mathematical thinking was contextual and embedded in action, rather than consciously used to achieve perfection or precision in the enlargement. Only the teacher in collaboration with the children can determine whether more formal mathematics inquiry would enrich or interrupt the investigation and evaluate the potential of such a direction to support the development of identity and power. In addition, we wonder whether the imperfection of the enlargement contributed to the children's view of their created object as art. If so, then movement to pursue disciplinary complexity by exploring and developing more precision in lettering could spoil the artistic character of the creation. Again, only Rachel and the children could evaluate the advantages and potential disadvantages of movement toward more precision in lettering using measurement to support new insights about "gap lengths" and their impact on the lettering. There is always a possibility that pursuing disciplinary complexity in such interactions will result in shifting priorities in a landscape of investigation in ways that could diminish the experiences of authority, autonomy, success, and relationships that derived from the project. Further research exploring characteristics of investigations, teachers, students, and systems of interaction may allow IdME researchers to hypothesize about and provide examples of when and how "mainly mono-disciplinary" (Williams et al., 2016, p. 19) investigations produce dilemmas, recognized by teachers as potentially mathematically productive, should be pursued.

Examples of IdME contexts have the potential to illustrate opportunities for learners, needs of teachers, and unearth researchable questions for the IdME research community. In the case of the landscape of investigation in the art class, we see opportunities to discuss dimensions of children's experience such projects afford.

As Engle (2011) points out, children's curiosity is endless. In the example of the art class children's curiosity was piqued, intuition about space developed, and the children used their voices in powerful ways to communicate about their work. The children redefined dimensions of experience in art, in ways that contrast with typical definitions in the discipline of mathematics. This example illustrates possibilities for the study of IdME contexts, in particular suggesting that one strand of study should involve inquiry into shifts in power and identity in such contexts.

The study of this art class suggests the potential of "mainly mono-disciplinary" (Williams et al., 2016, p. 19) contexts and actions of teachers for creating experiences that extend ways of operating, being, and knowing typically developed in the discipline of mathematics. The artistic investigation, of the children, generated opportunities for their reconceptualization of their identities as learners of mathematics. The art classroom environment afforded children with the chance to liberate themselves from self-images shaped by experiences of success or failure in the mathematics classroom. Repeated failure tends to leave students feeling disempowered when trying to learn mathematics. By shifting the learning of mathematics to other subject areas, teachers have a chance to rebuild students' self-confidence, self-image, and views of their power and human identity. The respect and solidarity that children showed towards each other in the wake of appreciating each other's artistic creations was evident in the conversations held throughout the project. We conjecture that when the authority of determining what is problematic, how to explore such problems, and what is right and wrong, is in the hands of the students, they are flexible in accepting alternative ways of seeing things and as a result, more accepting of the ideas and products of others. Constructing IdME contexts involves creating opportunities for students to explore, and represent creatively, their worlds, thereby reframing their identities and sense of power in school in ways that support their efforts to reach their human potential.

Rachel's work with the children and their engagement in the landscape of investigation involved the discipline of mathematics as it is embedded in the actions of creating objects for society. Yet without explicit discussion of mathematics, what are children learning? Contexts where mathematics is embedded in action as part of the "continuum of disciplinarity" (Williams et al., 2016, p. 32) is not identified and separated from other disciplines, yet as Cox (2013) has shown in the case of proportional reasoning, the intuition that develops in such context may be critical to success in mathematics class. Researchers have discussed human actions used in such contexts, and the danger in devaluing them (Civil, 2016; Fasheh, 2015). Human actions embedded in the beauty of creating in domains outside mathematics (Fasheh, 2015) may support individuals to reach their human potential (D'Ambrosio & D'Ambrosio, 2013). What is less understood is the value of mathematical intuition that develops when children are involved with ideas and peers in ways that support learner empowerment. Is such mathematics, viewed through the lens of the discipline, necessarily weak? Further research in IdME is needed to explore whether and how such context balance disciplinary goals and the primary goal of education as described by D'Ambrosio and D'Ambrosio. Additional narratives can illustrate ways teachers and students use such context to inform mathematical identities of young people and their sense of power to identify, communicate, and participate in constructing solutions to problems that contribute to self- and societal-satisfying activity.

We further wonder how multi-disciplinary teams of teachers, with a shared goal of helping students reach their human potential, might use "disciplinary 'identity' and division of labour" (Williams et al., 2016, p. 4) to contribute to this goal. Rachel sought to support her colleagues, other elementary school teachers, in their efforts to build children's critical thinking in mathematics. She felt satisfaction at contributing to colleagues' discipline specific goals and school culture through the children's production of a mural. Research focussed on multi-disciplinary teacher teams could take up this idea and explore teacher's emotional and affective responses to engagement in making curriculum to address discipline specific goals as well as to support engaging future citizens in understanding of the value of being human and contributing to the work and joy of other humans.

References

Civil, M. (2016). STEM learning research through a funds of knowledge lens. *Cultural Studies of Science Education, 11,* 41–59.

Clandinin, D. J., & Connelly, F. M. (2000). *Narrative inquiry*. San Francisco, CA: Jossey-Bass.

Clandinin, J. (2006). Narrative inquiry: A methodology for studying lived experiences. *Research Studies in Music Education, 27,* 44–54.

Cox, D. (2013). Similarity in middle school mathematics: At the crossroads of geometry and number. *Mathematical Thinking and Learning, 15,* 3–23.

D'Ambrosio, U. (1990). The role of mathematics education in building a democratic and just society. *For the Learning of Mathematics, 10*(3), 20–23.

D'Ambrosio, U., & D'Ambrosio, B. S. (2013). The role of ethnomathematics in curricular leadership in mathematics education. *Journal of Mathematics Education at Teachers' College, 4,* 10–16.

Engle, S. (2011). Children's need to know: Curiosity in schools. *Harvard Educational Review, 81,* 625–645.

Fasheh, M. (2015). *Over 68 years with mathematics: My story of healing from modern superstitions and reclaiming my sense of being and well-being.* Paper presented at the Eighth International Mathematics Education and Society Conference, Portland, Oregon.

Freire, P. (1998). *Pedagogy of freedom: Ethics, democracy, and civic courage*. Maryland: Rowman & Littlefield Publishers Inc.

Freire, P., D'Ambrosio, U., & Mendonça, M. C. (1997). A conversation with Paulo Freire. *For the Learning of Mathematics, 17*(3), 7–10.

Guitérrez, R. (2013). The sociopolitical turn in mathematics education. *Journal for Research in Mathematics Education, 44*(1), 37–68.

Gutstein, E., & Peterson, B. (2005). *Rethinking mathematics: Teaching social justice by the numbers*. Milwaukee: Rethinking Schools Ltd.

Paley, V. (1986). On listening to what the children say. *Harvard Educational Review, 56,* 122–131.

Paley, V. (2011). Voices inside schools: Getting to know Derek. *Harvard Educational Review, 81,* 745–750.

Polkinghorne, D. (1995). Narrative configuration in qualitative analysis. *International Journal of Qualitative Studies in Education, 8*(1), 5–23.

Skovsmose, O. (2001). Landscapes of investigations. *ZDM, 33,* 123–132.

Skovsmose, O., & Valero, P. (2001). Breaking political neutrality: The critical engagement of mathematics education with democracy. In B. Atweh, H. Forgasz, & B. Nebres (Eds.), *Sociocultural research on mathematics education: An international perspective* (pp. 37–55). Mahwah, NJ: LEA.

Stephens, C. (2013). *In the art room: The Andy Warhol mural*. Retrieved March 18, 2014, from http://cassiestephens.blogspot.com/2013/09/in-art-room-andy-warhol-mural.html.

Streefland, L. (1984). Search for the roots of ratio: Some thoughts on the long term learning process (towards…a theory) part I: Reflections on a teaching experiment. *Educational Studies in Mathematics, 15,* 327–348.

Walshaw, M. (2005). Getting political and unravelling layers of gendered mathematical identifications. *Cambridge Journal of Education, 35,* 19–34.

Williams, J., Roth, W.-M., Swanson, D., Doig, B., Groves, S., Omuvwie, M., et al. (2016). *Interdisciplinary mathematics education: State of the art*. Cham: Springer. https://doi.org/10.1007/978-3-319-42267-1.

Yackel, E., & Cobb, P. (1996). Sociomathematical norms, argumentation, and autonomy in mathematics. *Journal for Research in Mathematics Education, 27,* 458–477.

Chapter 15
Inter-disciplinary Mathematics: Old Wine in New Bottles?

Brian Doig and Wendy Jobling

Abstract This paper attempts to demonstrate that inter-disciplinary mathematics is an old practice, newly rediscovered, and formerly accessible to everyone, but problematised by modern times. Evidence of interdisciplinary mathematics, now often termed STEM, is presented from history and from more recent curriculum documents. Research into the benefits of integrated approaches to STEM education give qualified support to such approaches, and suggests characteristics defining effective interdisciplinary learning. An example of a project-based approach is examined for its contribution to thinking about how inter-disciplinary mathematics might be more generally applied to student learning in Primary and Secondary schools in modern times. Curricular considerations and examples are examined for inter-disciplinary possibilities, while some caveats are presented to temper any rush to inter-disciplinarity without due consideration of the consequences.

Keywords Mathematics · Interdisciplinarity · STEM · Caveats

15.1 Early Inter-disciplinarians and Interdisciplinarity

The term "Inter-disciplinary mathematics" is related to, or perhaps subsumes, terms such as polymath from the Greek: πολυμαθής, "having learned much" and scientist, coined from the Latin scientia "knowledge, a knowing, expertness" (On-line Dictionary of Etymology, 2015). Either way, we recognize Leonardo da Vinci and Galileo Galilei as polymaths, as inter-disciplinary thinkers. More recently, in the classic work by Bell (1951) Mathematics: Queen and Servant of Science, the nexus between mathematics and science was explicit.

Further, within educational curricula at pre-university levels, the use of mathematics in other discipline areas was taken for granted. For example, in the Australian context, the Victorian Education Department guidelines for teachers, *The method of*

B. Doig (✉) · W. Jobling
Deakin University, Melbourne, Australia
e-mail: Badoig666@deakinedu.au

B. Doig et al. (eds.), *Interdisciplinary Mathematics Education*,
ICME-13 Monographs, https://doi.org/10.1007/978-3-030-11066-6_15

teaching arithmetic (Department of Education, 1944), teachers of Grade VII were exhorted:

> children should actually participate in finding the size of rooms, areas of floors, and the amount of timber required them. The *co-operation of tradesmen could be secured to check the methods adopted* (p. 161) (our italics).

Clearly, it was assumed that tradesmen used mathematics, although their methods may not be, necessarily, those taught in schools. Ten years later, in the Course of Study for Primary Schools, Arithmetic, Grade VII, teachers in rural areas were again reminded that arithmetic must include "Practical investigations and exercises appropriate to the major primary industries of the locality (at least two of which must be taken" (Department of Education, 1954, p. 4). Of course, one can argue that trade mathematics, or even arithmetic, is not mathematics: but it is *part* of mathematics. Thus, tradesmen, it could be said, continue the tradition of the polymaths and early scientists.

15.2 Modern Times

The modern term 'inter-disciplinary mathematics', and related terms, such as STEM (Science, Technology, Engineering and Mathematics) and STEAM (Science, Technology, Engineering, Arts, and Mathematics), have become prominent in recent decades. A notable inclusion in this area is that of technology, often taken to be a reference to digital technologies, rather than technology in its broader sense. This broader sense is, for example, that "technology is the term that includes all the technologies developed and used by people in the purposeful application of knowledge, experience, and resources to create products and processes that meet human needs" (Australian Education Council, 1992).

In Western Australia, the Education Department took the step of defining its technology process strand as designing, making, and appraising (DMA) (Education Department of Western Australia, 1994), to make the study of technology an authentic, real life practice for the students. This meant that technology in Western Australian schools included metal-work, carpentry, jewellery making, pottery, cooking, and even horsemanship.

In 2003 Mason, Mittag, and Taylor, prompted by calls from national mathematics (National Council of Teachers of Mathematics, 2000); and science associations (American Association for the Advancement of Science, 1989), published *Integrating mathematics, science, and technology*. However, while their work was replete with mathematics and science, the technology was restricted to digital technology, effectively only graphic calculators.

15.2.1 Integrative Approaches to Inter-disciplinary Learning

Disciplinary Learning

Design Sciences, which, at that time in Greece, he claimed, included architecture, engineering, and even medicine and economics: this was from a perspective that "the mission of the Design Sciences is the design and manufacture of artificial objects, having certain desirable properties" (p. 134). In order to emphasise his claim for the importance of mathematics in the Design Sciences, he took a case from architecture "where the roofs of buildings with wide openings (closed stages, swimming pools etc.) are usually designed in the Voskoglou (2006) argued that mathematics was important for the form of a saddle, because such type of surfaces have big resistance to bending" (p. 134). This saddle is a hyperbolic paraboloid. A further example cited was that of Markov chains, "a successful combination of Linear Algebra and Probability theory … [in] Operations Research" (p. 137). Of course, such high levels of integration of the mathematics into the Design Sciences may mean that the mathematics may lose its identity and become simply part of that particular Design Science.

According to Becker and Park (2011) "Science, technology, engineering, and mathematics (STEM) education is a crucial issue in current educational trends" (p. 23), but, that "due to the lack of a comprehensive review regarding the effects of integrative approaches among STEM subjects on academic achievement, many teachers are unaware of the benefits of the integrative approaches for student learning" (p. 24). Integrative approaches are defined by Wells and Ernst (2015) cited by Hayward (2016, p. 15) as technological/engineering design-based learning approaches that intentionally integrate the concepts and practices of science, and, or, mathematics education, with the concepts and practices of technology and, or, engineering education. Integrative STEM education may be enhanced through further integration with other school subjects, such as language arts, social studies, art, etc. (Sanders & Wells, 2010).

Note that this definition (clearly and very intentionally) excludes pedagogical approaches that do not situate the teaching and learning of STEM concepts and practices in the context of technological/engineering design-based activity. Furthermore, only technologies that are integral to the doing of designing/making/engineering constitute the T/E component in this definition. For example, simply employing instructional technologies to teach S/M concepts does not constitute the T/E component essential to integrative STEM instruction. According to Laboy-Rush (2011)

> [T]hrough an integrated approach to STEM education focused [*sic*] on real-world, authentic problems, students learn to reflect on the problem-solving process. Research tells us that students learn best when encouraged to construct their own knowledge of the world around them … [and it] is through integrated STEM projects that this type of learning can occur (p. 1).

However, is this claim substantiated? Becker and Park conducted a meta-analysis on some twenty-eight earlier studies, and produced effect sizes for each study. The results were mixed, with some studies showing large effect sizes for some integrative approaches, and lower effects for others. However, "Students who were exposed to integrative approaches demonstrated greater achievement in STEM subjects" (Becker & Park, 2011, p. 31). An interesting finding was that including mathematics in the mix of integrated subjects lowered effect sizes: "the effect sizes of students' achievement were small when mathematics was integrated" (p. 31).

Laboy-Rush (2011) cited Diaz and King (2007), who had suggested five characteristics of effective inter-disciplinary (STEM) projects for student learning. In essence, these were that students:

- have a variety of learning tasks to involve them in the learning process;
- receive explicit communications and explanations;
- have opportunities to model solutions, practise solving problems, and receive constructive feedback;
- engage in a student-centred instructional environment that focuses on their interests and needs; and
- receive support for their learning needs.

A recent example of a middle school integrated approach is that of the Prairie River Middle School, in the USA state of Wisconsin. There, the 'technology engineering' teacher had a group of students work together to build, from timber, a four metre rowing boat, that was raffled to raise money for charity (Jettinghoff, 2016), "Middle School students use project-based learning to improve math [*sic*] skills and overall academic achievement" (Alexandria Seaport, 2016). Other enterprises involving the building of boats appear to be a popular approach to the integration of mathematics and other STEM subjects.

15.2.2 Integration of STEM

A long-standing Victorian STEM competition, involving Primary- and Secondary-aged students in building and racing solar-powered model cars and boats, is an example of an integrated approach to STEM education.

The Victorian Model Solar Vehicle Challenge (MSVC) has been running for over twenty years and is a competition in which groups of students design, make, and race model solar vehicles: these are either wheeled vehicles or boats. In addition, entrants need to provide a poster communicating the building processes: the posters are part of the competition and are judged before being displayed at the challenge event. See www.modelsolar.org.au for more details and video of previous entries. The challenge is conducted at state and national levels, and there is the opportunity for winners to compete internationally too. For example, in the Victorian 2013 challenge, a team from Taiwan competed.

An important aspect of the MSVC is the link to curriculum areas and the integration of them. The main curriculum areas addressed in the MSVC are science, particularly the physical sciences, in which electrical circuits, solar production of electricity, friction, and air or water resistance, are important aspects. Appraise (DMA) process

Fig. 15.1 Model solar boats racing. Photograph: W. Jobling

is critical to producing an efficient model with the equipment and materials available. Testing of proto-types involves data collection and interpretation, as well as drawing logical meanings from the data for improvement to the model. Despite this, the rôle of mathematics is mainly in measuring. The model must keep within the required dimensions, and measurements must be taken during model testing and these data displayed on the poster, as well as interpreted for model improvement.

In terms of technology, the use of the iterative Design, Make, and Appraise (DMA) process is critical to producing an efficient model with the equipment and materials available. Testing of proto-types involves data collection and interpretation, as well as drawing logical meanings from the data for improvement to the model. Despite this, the rôle of mathematics is mainly in measuring. The model must keep within the required dimensions, and measurements must be taken during model testing and these data displayed on the poster, as well as interpreted for model improvement (Fig. 15.1).

Engineering, particularly electrical and mechanical engineering, clearly plays a large rôle in constructing a vehicle that moves by wheels or propellers, and is powered by solar cells. For the model solar boat challenge, in particular, hull shape and the propulsion system are major aspects of the need for some engineering understanding.

Communicating the processes and outcomes of the project, on a poster, helps to draw student attention to their journey to a finished model, as well as how to explain, to a wider audience, the processes and results of STEM projects, and their own learning from the experience.

Is the Solar Challenge project-based learning? If it is, then Donnelly's (2015) question, "Should we 'teach' interdisciplinarity at school?" (p. 3) would be answered in the affirmative. He suggests that project-based learning would teach students the inter-disciplinary links necessary for inter-disciplinary approaches at University, or at work.

> Making these links explicit to students can only be of benefit … [and] be of benefit to teachers too, working collaboratively across disciplines, sharing knowledge and experiences of pedagogical approaches and joint planning (p. 3).

Apparently, the Singapore Ministry of Education also thinks that inter-disciplinarity across the STEM disciplines is possible, as it announced in 2015 that 42 Secondary schools offer the Science, Technology, Engineering, and Mathematics Applied Learning Programme (STEM-ALP), and that by 2017 half of the 124 mainstream Secondary schools in Singapore would offer the programme.

The skills and competencies to be developed include:

- Scientific inquiry and literacy;
- Reasoning and problem solving;
- Design thinking;
- Computational thinking; and
- Data analysis and the use of technology (Ministry of Education, 2015, p. 4).

While Singapore is not alone in treading the STEM path, Singapore's noted achievements in international academic 'contests', such as TIMSS, suggests that the outcomes of this foray into STEM education through the Applied Learning Programme will be worth watching.

15.3 Caveats

While most of the STEM literature supports some form of integration between the STEM subjects, there is always the possibilities that some integrations may not contribute successful mathematical learning for the students (Becker & Park, 2011); Doig, Groves, and Williams (1996) reported on the mathematization of a science modelling activity with Primary school children aged between 10 and 11 years. The activity involved dropping a 'timer ball' to discover the height from which to drop an object for a falling time of one second. The children experimented progressively with a 0.25 s drop, a 0.5 s drop, and a 0.75 s drop. These data were then graphed and the children noted that the changes in height necessary for a longer falling duration were not linear. For example "Bob said, 'a quarter second is less than half of half a second, so if it worked the same then a whole second should be more than double half a second" (Doig et al. 1996, p. 4). Further, two high ability 11 year-olds "spontaneously found the differences between the distances [fallen in each quarter second] and concluded that the ball was accelerating' (p. 6). Later, with a group of 10 year-olds, the authors were unsuccessful in convincing the students that this was in fact the case. Thus, Doig et al. noted, that for some children "[t]he process seems almost circular at times—the data obtained from the practical activity is intended to inform the construction of a model to explain the observed behaviour, yet in order to interpret the data we need to view the data through the window of our existing models" (p. 6).

The Doig et al. example of acceleration, would suggest that inter-disciplinarity may not be a simple matter, and that a re-thinking of curriculum content and sequence may well be needed. In the example above, some preliminary science addressing gravity may have been helpful for the acceleration activity.

A further caveat is the rise of proponents for STEAM, where the 'A' stands for the Arts. In this, we find advocates theorizing that STEAM has benefits because "[a]dding the arts into the STEM equation can re-invigorate the platform, providing not only an interesting approach, but also the opportunities for the self-expression and personal connection new generations crave" (Land, 2013, p. 548). While it is sensible for proponents of the Arts to jump on the STEM bandwagon, claims such as Land's are not easily supported by evidence. For example, Madden et al. (2013) list several innovative integrated tertiary curricula starting up at the time (2013). However, despite the goodwill and efforts of those concerned, claims about the outcomes of courses in creativity integrated into the STEM area are not supported by evidence as yet. Time may yet prove the creators of these programmes to be prescient, but for now, one might say that STEAM is nothing but hot air. Newer calls for other disciplines to be integrated into STEM include religion (STREAM) and history (SHTREAM), and geography, but, at that stage, the acronym becomes unpronounceable!

In some countries, such as Australia, Primary teachers teach all subjects, and the full range of disciplines could be integrated, fully, in theory. However, Australian Primary teachers tend to teach in the silos of discipline areas as much as do Secondary teachers. This is an area needing research, as Primary teachers may be the best placed to adopt inter-disciplinary teaching.

Further, Mercedes-Benz provides an exemplary warning to those urging for, or employing, interdisciplinarity. In the mid-nineteen nineties Mercedes-Benz was keen to develop a car that would be "aerodynamic, safe, efficient, and maneuverable [*sic*]" (Buehler, 2015, p. 1). These criteria led designers and engineers to look to Nature for a possible solution. This, they thought, was to be found in the Boxfish (Ostraciidae Tetrodontiformes), which had remarkable capabilities. For, although box-like as its name suggests, the Boxfish was thought to have excellent hydrodynamic characteristics (low drag), a spacious body, and good stability. Further, the "carapace supposedly had unique, inherent, self-correcting stabilization properties" (Buehler, 2015, p. 1).

Thus, the Mercedes-Benz designers and engineers set out to create the Bionic concept car.

But, researchers at the Universities of Antwerp and Groningen, and the University of California (Los Angeles) reported that the Boxfish shape did not have lower drag, nor did its shape promote stability, but rather, the Boxfish used its inherent instability for fast manoeuvring (Van Wassenberg, van Manen, Marcroft, Alfaro, & Stamhuis 2015).

(For details of the Yellow box-fish, see https://australianmuseum.net.au/yellow-boxfish-ostracion-cubicus).

The lesson here is that a little knowledge is a dangerous thing, and one needs to engage with those with real expertise in the particular field. An inter-disciplinary team, rather than everyone a polymath?

The rise, and fall, of the *Sloyd* movement in the late nineteenth and early twentieth century has similarities to the sudden international support for STEM. While *sloyd* still has supporters (see, for example, Noe, 2016), it is hard to imagine how this Scandinavian idea spread like wild-fire internationally. According to Hoffman (1892) the word *sloyd* is derived from the Swedish word *slöjd*, which translates as crafts, handicraft, or handiwork and referred primarily to woodwork. The founder, Otto Salomon, with the financial support of his uncle, started a school for teachers in Nääs in the 1870s. The school attracted students from throughout the world and was active until around 1960.

Educational *sloyd*'s purpose was formative, as it was thought that the benefits of handicrafts in education built the character of the child, encouraging moral behaviour, greater intelligence, and industriousness. *Sloyd* had a major impact on the early development of manual training, manual arts, industrial education and technical education. But, today, in most parts of the world, *sloyd* is long forgotten. Could the "STEMmania" (Sanders, 2009, p. 20) suffer a similar fate? To those involved in the *sloyd* revolution just over a hundred years ago, the answer would most likely be 'yes'. So what stopped *sloyd*?

The opposition to *sloyd* argued that the *sloyd* approach was too rigid, and left no place for creativity, and, like other waves of educational change (for example, the so-called "new maths" of the 1960s) excitement and enthusiasm waned. Salomon argued that learning one craft, wood-working, well, was better than a cursory knowledge of many crafts. Is this a warning for the longevity of STEM?

15.4 Discussion

While claims that mathematics supports human endeavours in science, engineering, and technology are beyond dispute, the implications for educational practice are not so clear. In the curriculum materials presented here, half a century ago mathematics was seen as the tool for other aspects of students' lives. Recently however, this has disappeared, and the rise of calls for STEM education underscore this. However, the question for curriculum developers, teachers, and students is how to install inter-disciplinary mathematics into the fabric of the modern school. Extra-curricular ideas, such as the Victorian Model Solar Vehicle Challenge may point the way for those teachers dedicated enough to make their time and expertise available, but is it sufficient that only a small proportion of students are engaged?

Perhaps, support for a wider audience of teachers to take up inter-disciplinary mathematics with their students lies in more texts such as that of Mason, Mittag, and Taylor (2003), or with projects like the University of California Berkeley's Lawrence Hall of Science's *Science Education for Public Understanding Program* (SEPUP) (2015) materials.

The Ontario, Canada, Ministry of Education introduced an Integrated Curriculum in 2006, and Drake and Reid (2010) maintain that a similar integrated curriculum had existed in Ontario in the 1930s (p. 1). In such integrated curriculum approaches, they claim, "students … demonstrate academic performance equal to, or better than, students in discipline-based programs. In addition, students are more engaged in school, and less prone to attendance and behaviour problems" (p. 1). Note though, that this is an integration of all school subjects, and is not restricted to STEM subjects. Results were most encouraging, as "[t]eachers and administrators identified student engagement as the most positive aspect of integration … Strong engagement levels alleviated behaviour problems" (Drake & Reid, 2010, p. 3). Further, "[t]eachers, impressed by the level of classroom discussion, concluded, 'integrated curriculum lends itself to higher order thinking skills" (Drake & Reid, 2010, p. 3). However, whether these observations carry through to greater understanding and skills in the integrated subjects is yet unproved.

But, are teachers ready to undertake inter-disciplinary mathematics? Or science? The answer to this question most likely lies within the universities who educate teachers of both mathematics and the other STEM disciplines. An example, of one university approach, is that of the Aggie-Center at the Texas A&M University, where Summer Camps for teachers interested in inter-disciplinary STEM teaching are held each year (Aggie-STEM, 2015). These are well supported and numbers of applicants outruns the number of available places each year.

We believe that we have shown, the path to successful inter-disciplinarity is not without its twists and turns. However, history tells us that the interdisciplinarity of the past was accepted as the norm, however moderate we may now think its level. Perhaps the path to success lies not in more subject areas, nor in teacher specialization, but rather in fewer, key, subject areas coupled with every teacher being an inter-disciplinary teacher. Project work that engages students, and teachers, would be an essential aspect of such an inter-disciplinary approach, so perhaps the saying 'Old wine in new bottles' is apt!

However, as with many educational 'carts', one should consider the 'horse' first. That is to say, that some resolution of the 'why STEM' question might assist educators, and administrators, to think clearly about what might constitute inter-disciplinarity in their context. Even if a project-based approach were taken, which projects would satisfy the 'why STEM' question. These, and other questions, need to be considered before STEM is over-run with vested interests, rather than educational needs, which has been the fate of other innovations, the digital innovation being a current prime example.

References

Aggie-STEM. (2015). *Aggie-STEM summer camp*. Retrieved from:http://aggiestem.tamu.edu/resources/student-resources/summer-camp.

Alexandria Seaport Foundation. (2016). *Middle school math.* Retrieved from: http://alexandriaseaport.org/.

American Association for the Advancement of Science. (1989). *Science for all Americans: Project 2061* (Author Ed.). Washington, DC.

Australian Education Council. (1992). *Technology-a curriculum profile for Australian schools.* Carlton, Victoria: Curriculum Corporation.

Becker, K., & Park, K. (2011). Effects of integrative approaches among science, technology, engineering, and mathematics (STEM) subjects on students' learning: A preliminary meta-analysis. *Journal of STEM Education, 12*(5 & 6), 23–36.

Bell, E.T. (1951). *Mathematics: Queen and servant of science.* London: G. Bell and Sons.

Buehler, J. (2015). *A real drag.* Articles. Retrieved March 23, 2016 from http://www.slate.com/articles/health_and_science/science/2015/03/mercedes_benz_bionic_car_boxfish_stability_and_agility_paradox_finally_solved.html.

Department of Education. (1944). *The method of teaching arithmetic.* Melbourne: Victorian State Government.

Department of Education. (1954). *Course of Study for Primary Schools.* Melbourne: Victorian State Government.

Diaz, D., & King, P. (2007). *Adapting a post-secondary instructional model to K-5 mathematics instruction.* Clemson: Clemson University.

Doig, B., Groves, S., & Williams, J. (1996). Mathematization: Support or hindrance in young children's science? In: *Proceedings of the Eighth International Congress on Mathematical Education.* Seville, Spain.

Donnelly, R. (Wednesday, 18 November, 2015). Should we 'teach' interdisciplinarity at school? *The BERA Blog: Research matters.* Retrieved from: https://www.bera.ac.uk/blog/should-we-teach-interdisciplinarity-at-school?/BERA.

Drake, S., & Reid, J. (2010). Integrated curriculum. *What works? Research into Practice,* (28). Retrieved from:http://www.edu.gov.on.ca/eng/literacynumeracy/publications.htm.

Education Department of Western Australia. (1994). *Technology and enterprise learning area.* Perth: Education Department of Western Australia.

Hayward, J. (2016). *An analysis of secondary integrated STEM lesson plans: Common characteristics, learning expectations and the impact from the teacher's definition of I-STEM.* Curriculum and Instruction, University of Arkansas. PhD

Hoffman, B. (1892). *The sloyd system of wood working* (B. B. Hoffman, Trans.). New York: American Book Company.

Jettinghoff, R. (2016). Launchings. *Wooden boat,* 87.

Laboy-Rush, D. (2011). *Whitepaper: Integrated STEM education through project-based learning.* Retrieved from:www.learning.com/stem/whitepaper.

Land, M. (2013). Full STEAM ahead: The benefits of integrating the arts into STEM. *Procedia Computer Science, 20,* 547–552.

Lawrence Hall of Science. (2015*). Science education for public understanding program (SEPUP).* Retrieved from: www.sepuplhs.org/.

Madden, M., Baxter, M., Beauchamp, H., Bouchard, K., Habermas, D., Huff, M., et al. (2013). Rethinking STEM education: An interdisciplinary STEAM curriculum. *Procedia Computer Science, 20,* 541–546.

Mason, D., Mittag, K., & Taylor, S. (2003). *Integrating mathematics, science, and technology.* Boston: Pearson Education.

Ministry of Education, Singapore. (2015). *42 Secondary schools offering science, technology, engineering, and mathematics applied learning programme (STEM ALP).* Press releases. Retrieved from: http://www.moe.gov.sg/media/press/2015.

National Council of Teachers of Mathematics. (2000). *Principles and standards for school mathematics.* Reston, VA: The National Council of Teachers of Mathematics.

Noe, R. (2016). *Sloyd education theory: Making things with your hands makes you smarter*. Retrieved from: http://www.core77.com/posts/58789/Sloyd-Education-Theory-Making-Things-With-Your-Hands-Makes-You-Smarter.

On-line Dictionary of Etymology. (2015). *On-line dictionary of etymology*. Retrieved from: http://www.etymonline.com.

Sanders, M. (2009). STEM, STEM education, STEMmania. *The technology teacher*, December/January 2009, pp 20–26.

Sanders, M., & Wells, J. (2010). *Integrative STEM education*. Retrieved from: www.soe.vt.edu/istemed.

Van Wassenbergh, S., van Manen, K., Marcroft, T., Alfaro, M., & Stamhuis, E. (2015). Boxfish swimming paradox resolved: Forces by the flow of water around the body promote manoeuvrability. *Journal of the Royal Society: Interface, 12*(103). https://dx.doi.org/10.1098/rsif.2014.1146.

Voskoglou, M. (2006, July 9–15). *The importance of mathematics for the design sciences*. Paper presented at the Commission Internationale pour l'Etude et l'Amelioration de l'Enseignment des Mathematiques, Srni, Czech Republic.

Part IV
Teacher Education and Teacher Development

Rita Borromeo Ferri

Chapter 16
Teacher Education and Teacher Development

Rita Borromeo Ferri

Abstract The two chapters on teacher education for Interdisciplinary Mathematics Education are introduced. It is argued that teacher education is required for the innovations that interdisciplinarity demands, and that the success of interdisciplinary education stands or falls on the state of teacher preparation for it.

Keywords Teacher education · STEM education · Interdisciplinary teaching

Promoting STEM education is a main goal of educational policy in many countries worldwide. The scientific and technological progress, in commercial countries will persist, and thus people are needed who are qualified in Science, Technology, Engineering and Mathematics—the STEM fields. This is at the behest of companies, who have problems in recruiting, highly qualified engineers, and IT specialists, for example, and recently it has become a clearly defined educational goal, that STEM-programmes are supported by governments. In the United States, the initiative for STEM has a longer tradition (National Research Council, 2014) than in many other countries, although the interest of their students in the STEM-fields still remains rare. In this context the Obama administration announced the 2009 "Educate to Innovate" campaign to motivate and inspire students to excel in STEM subjects.

Similarly, results of a recently published report of the *status quo* of STEM-education in Europe, supported by the European Union (Galev, 2015), investigating teachers, students, and experts from industry, show that there has been some progress, but STEM-education in school is still taught mainly more theory-orientated than practice-orientated, within separated disciplines. One reason, among others, could be the fact that there is no systematic teacher education in this field in many European countries. The implementation of an interdisciplinary approach, from kindergarten age to school, and university, levels, is still a challenge for teachers, and for those who are educating the teachers. Interdisciplinary learning and teaching require, on the one hand, well-prepared teachers, and on the other hand, adequate teaching materials for every-day lessons in school (not to speak of curriculum and assessment regimes).

R. Borromeo Ferri (✉)
University of Kassel, Kassel, Germany
e-mail: borromeo@mathematik.uni-kassel.de

B. Doig et al. (eds.), *Interdisciplinary Mathematics Education*,
ICME-13 Monographs, https://doi.org/10.1007/978-3-030-11066-6_16

In the following two articles, the reader gets an insight into good teacher education approaches for interdisciplinary teaching, and learning, in several countries. In particular, it becomes clear, how important collaborations between teachers from different disciplines are, and furthermore, how vital it is to make the effects of this interdisciplinary teaching explicit, and transparent, to colleagues.

The first article, by Bobořová, Čeretková, Tirpáková, and Markechová, shows the inclusion of the interdisciplinary approach in Biology Teacher Trainees in Slovakia. The researchers comment that there has been almost no significant progress in incorporating interdisciplinary approaches into teacher education curricula at universities in their country. This results from the fact that the curriculum in Slovakia is separated into Mathematics, Biology, Chemistry, Physics and Computer Science. In order to connect these fields, their study was conducted with the aim to develop modern teaching material for biology teacher trainees with respect to the interdisciplinarity of mathematic and biology. Furthermore, the research group focused on analysing the key mathematical competencies for biology teacher trainees, with respect to the requirements of modern biology education. Finally, the researchers received valuable, authentic, feedback from teacher trainees, to help university educators identify the areas of strength, and weakness, of the materials used in order to improve interdisciplinary teaching.

The second paper, by Wilhelm and Fisher, from the United States of America, with a focus on creating academic teacher scholars in STEM education by preparing preservice teachers as researchers. The authors describe what Research Experiences for Undergraduates (REU) Fellows reported, regarding their experiences within a research-intensive programme in STEM Education. One main result of their study, was, that the Fellows most frequently noted the increase of their interpersonal collaborations with other future STEM teacher researchers. A characteristic, of the research, is the fact that it is the first to examine the effectiveness of an academic year long interdisciplinary STEM Education REU programme.

Both papers show that there is a lot of progress in STEM teacher education in different parts of the world. It becomes clear, that well-educated teachers for interdisciplinary teaching, who are trained in methods and content, are much needed. As well as these, and of other competencies, one aspect should be added, which we assume implicitly: we need teachers, who are open-minded enough not to see only their own favoured subject, or discipline, but who like to connect several disciplines, discuss their links with colleagues, create ideas, and make interdisciplinary teaching, and learning lively and motivating for their students.

References

Galev, T. (2015). Status Quo der MINT-Bildung in Europa: Ergebnisse der MARCH Studie. *Projekt Lebenslanges Lernen*. Europäische Union.

National Research Council. (2014). *STEM learning is everywhere: Summary of a convocation on building learning systems*. Washington, DC: The National Academies Press.

Chapter 17
Inclusion of Interdisciplinary Approach in the Mathematics Education of Biology Trainee Teachers in Slovakia

Ivana Boboňová, Soňa Čeretková, Anna Tirpáková and Dagmar Markechová

Abstract In order to eliminate the lack of teaching materials relevant to interdisciplinary teacher education, we implemented a worksheet about human blood. Altogether 64 students of bachelor and master study programmes of biology teacher training at Constantine the Philosopher University (CPU) in Nitra, Slovakia, filled in a 12-item worksheet which required the application of mathematical methods for answering biology-related questions. Moreover, students were asked to fill out a questionnaire regarding their attitudes to this material. By means of a content analysis of the teacher trainees' performance on each of the 12 items, their current mathematical competencies (unit conversion, calculation of percentages, the rule of three, and combinatorics) were evaluated. Based on the statistical evaluation we conclude that the students' success rate in tasks focused on unit conversion, calculation of percentages, and the rule of three, was significantly different than in tasks focused on combinatorics. This suggests that the higher knowledge shown in Mendelian concepts (combinatorics) is probably based on studying it in an interdisciplinary manner in their general biology courses. Finally, we conclude that the analysis of feedback received from students provides the university educators with the opportunity to adjust the content of the worksheet, and to improve their teaching strategies, in interdisciplinary contexts.

Keywords Interdisciplinarity · Mathematics · Biology · Teacher trainees · Mathematical competence

I. Boboňová (✉) · S. Čeretková · A. Tirpáková · D. Markechová
Constantine the Philosopher University, Nitra, Slovakia
e-mail: ibobonova@ukf.sk

S. Čeretková
e-mail: sceretkova@ukf.sk

A. Tirpáková
e-mail: atirpakova@ukf.sk

D. Markechová
e-mail: dmarkechova@ukf.sk

B. Doig et al. (eds.), *Interdisciplinary Mathematics Education*,
ICME-13 Monographs, https://doi.org/10.1007/978-3-030-11066-6_17

17.1 Introduction

Over the past several years, an increasing enthusiasm has been observed for teaching approaches that combine mathematics and biology. Many researchers have discussed the need for increasing interdisciplinary teaching of these disciplines (Bialek & Botstein, 2004; Brent, 2004; Gross, Brent, & Hoy, 2004; May, 2004; Pevzner & Shamir, 2009; Pursell, 2009; Steitz, 2003; Tra & Evans, 2010). Consequently, the call for integrating more quantitative work in biology education has led to the production of new teaching tools that improve quantitative skills. Different strategies have been proposed that educators, and institutions, could use to increase the mathematical competence of biology students. These strategies include a greater integration of mathematical problems into science classes (Hester, Buxner, Elfring, & Nagy, 2014; Hodgson et al., 2005; National Research Council [henceforth NRC], 2003), more frequent use of biological context in traditional mathematics courses (NRC, 2003; Robeva & Laubenbacher, 2009), and the development of biology textbooks that include quantitative problems, or computational exercises (Jungck, 2005; Madlung, Bremer, Himelblau, & Tullis, 2011).

Despite the prevalence of the subject-centred curriculum in Slovak public secondary schools, in the last decade, interdisciplinary approaches have been implemented, also, in secondary education in Slovakia, especially through several regional, as well as international, educational and research projects, such as Compass, DynaMat, Primas, and recently EU LLP Comenius project MaT^{2}SMc (Čeretková, Jakab, & Naštická, 2015; Melušová, Šunderlík, & Čeretková, 2012; Pavlovičová, Rumanová, & Švecová, 2013; Sandanusová, 2006; Ulovec, Čeretková, Dockerty, Molnár, & Spagnolo, 2009). However, there has been no significant progress in incorporating interdisciplinary approaches into teacher education curricula at universities in Slovakia. In other words, although considerable amount of attention has been paid to an interdisciplinary approach in secondary education, the institutional education of future teachers lacks a particular focus on interdisciplinary issues.

Teaching through such a new interdisciplinary perspective requires appropriate adjustments to approaches and pedagogies (Jungck, Gaff, Fagen, & Labov, 2010; NRC, 2003, 2008). In order to ensure the sustainable success of such interdisciplinary work, it is important to make mathematical approaches an integral component of biology teacher education at universities. Since teachers are one of the most important factors in primary, and secondary, education (Kalin & Zuljan, 2007; Šorgo, 2010), any new trend should be introduced, most importantly, to teachers, who, consequently, can be expected to pass their general, and interdisciplinary knowledge, and skills, to students. Moreover, the beginning of a teaching career is in some ways the period that is associated with a certain shock, because the teacher trainee, who used to receive mostly theoretical training for the profession, becomes a teacher, who is exposed to a variety of educational situations every day. The basis of the pedagogical skills of a novice teacher is their experience: from these experiences a novice teacher chooses to use the situation that occurs in practice (Viteckova et al., 2016). Thus, the safest way to achieve success in interlinking a certain couple of disciplines, e.g. biology and mathematics, it seems that the proper development, of the professional development of pedagogical mathematical and biological content knowledge in the pre-service education of future teachers, should be based on effective teaching practice centred on active interdisciplinary learning techniques.

Based on these facts, the present study aims (1) to develop modern teaching material for biology teacher trainees, with respect to the interdisciplinarity of mathematics and biology, in use in the educational processes at Constantine the Philosopher University in Nitra, Slovakia (henceforth CPU), (2) to analyse the key mathematical competencies of biology teacher trainees with respect to the requirements of modern biology education, and (3) to receive valuable, and authentic, feedback from teacher trainees to help university educators identify the areas of strengths, and weaknesses, of the applied worksheet, in order to improve this interdisciplinary teaching material.

17.2 Methods and Instruments

17.2.1 Development of Teaching Material

Looking through the existing teaching and learning materials in Slovakia, we soon discovered that these materials were designed mostly by biology, or mathematics educators, and that they did not offer active support, or sufficient opportunities, for collaboration. Being an interdisciplinary team of mathematics and biology teacher educators, we set out to improve the situation. In order to eliminate the lack of teaching materials, relevant to interdisciplinary teacher education, we implemented a material about human blood for biology teacher trainees, with respect to the interdisciplinarity of mathematics and biology, with the perspective of further use in educational process at CPU. The developed material contains the lesson description, and a 12-item worksheet, that requires application of mathematical methods (unit conversion, calculation of percentages, the rule of three, and combinatorics) for answering biology-related questions at the secondary school level.

Following Choi and Pak (2006), Stock and Burton (2011), and Williams et al. (2016), we recognize various levels of the term interdisciplinarity which, we use in its broadest sense.

First, multi-disciplinarity is characterized as the least integrative from all integrated approaches yet, equally, it is arguably the most attainable. Multi-disciplinarity draws on knowledge from several disciplines, but stays within the boundaries of those fields. There is no attempt to cross boundaries, or generate new integrative knowledge. For example, mathematics can be utilized in other subjects for the benefit of learning mathematics, or for the benefit of learning the other subject, using mathematics as a tool, or a generalisation. The connection between subjects can be a common element between them, such as the use of length units in mathematics and biology (Fig. 17.1).

Second, interdisciplinarity, in its narrow sense, involves the interaction between two, or more, (different) disciplines, and occurs at the interface between the disciplines. This may range from the sharing of ideas, to full integration of concepts, methodology, procedures, theory, terminology, data, and organization of research (Fig. 17.2).

Third, trans-disciplinarity is probably the most desirable, and yet difficult, to obtain form of integrated approach. Trans-disciplinarity is a specific form of interdisciplinarity in which boundaries between, and beyond, disciplines are transcended, and knowledge and perspectives from different scientific disciplines, as well as non-scientific sources, are integrated (new disciplines and theories are synthesized). Trans-disciplinary approaches can be exemplified in Mendel's laws of inheritance, which are based on mathematical probability and combinatorics (Fig. 17.3).

As shown in the examples above, (Figs. 17.1, 17.2 and 17.3) the analysed teaching material for biology teacher trainees, comprised tasks with various levels of interdisciplinarity (in the broad sense). In some of the tasks, the relation of biology and mathematics was rather loose (Fig. 17.1), which we consider to be a case of multi-disciplinarity. In other tasks, the connection of the two disciplines is obvi-

A typical human erythrocyte has a biconcave shape with disc diameter approximately 7.5μm and a thickness at the thickest point 2.1μm and a minimum thickness in the centre 1μm, being much smaller than most human cells.

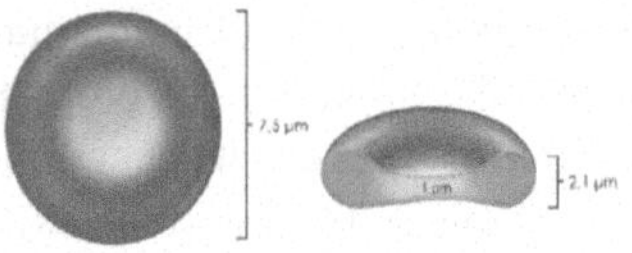

Task 1: *Convert the length values of erythrocyte from micrometres to the most commonly used units of length, millimetres and metres.*

Fig. 17.1 Size of erythrocyte

Erythrocyte has a biconcave shape and an average volume of about 90 femtolitres with a surface of about 136 square micrometres.

Task 2: *Calculate, as a percentage, how much the surface area of the biconcave shape of erythrocytes is bigger than hypothetical spherical cells having the same volume. Why is it important to consider the surface area of erythrocytes?*

Fig. 17.2 Geometric properties of erythrocyte

Task 3: *Parents have four children. Is it possible that each child has different ABO blood type? Explain.*

Fig. 17.3 ABO blood-group genetics

ously closer (Fig. 17.2), which we refer to as interdisciplinarity in a narrow sense. In trans-disciplinary issues, the gap between the disciplines diminishes, and the relation between them is very tight (Fig. 17.3).

17.2.2 Mathematical Competencies Assessment

Biology teacher trainees at CPU have a very diverse set of mathematical competencies from their previous secondary education. Moreover, they are not required to take mathematical courses during their study at the university, unless their second major is mathematics.[1] Therefore, the study described in this paper, aims at assessing the mathematical competencies of university students intending to become biology teachers. Altogether 64 students, of bachelor and master study programmes of biology teacher training at CPU, filled in a 12-item worksheet, that required application of mathematical methods (unit conversion, calculation of percentages, the rule of three, and combinatorics) for answering biology-related questions. None of the biology teacher trainees studied mathematics as their second major. Teacher trainees could obtain scores ranging from 1 to 5 for each of the items, where 1 = excellent; 2 = very good; 3 = good; 4 = fair and 5 = poor. Students' answers were evaluated statistically, which allowed determination and comparison of the rates of students' success. A non-parametric Friedman test was conducted to evaluate if there was any significant difference between the average scores achieved by biology teacher trainees in their solutions with respect to the four mathematical methods required by the tasks. Once a significant difference was found, a Neményi test was performed to find where the difference actually occurs (Anděl, 2003). The software STATISTICA was used for the data analysis, and all the tests were done at a significance level of 0.05.

17.2.3 Feedback Obtained in Questionnaire Survey

After the interdisciplinary worksheet implementation, participants were asked to fill in an anonymous questionnaire regarding their attitudes to the implemented teaching material. Teacher trainees were asked to evaluate the theoretical biology introduction and parts of the practical computational exercises. By means of a 5-point Likert scale they expressed the level of their agreement with five statements regarding the following areas: comprehensibility, interest, curricular requirements, teachers' practical requirements, and collaboration of teachers. In addition, teacher trainees were asked about their personal opinion of this material. The data obtained were summarized by means of descriptive statistics.

[1] In Slovakia, teacher trainees, for secondary education, study two majors simultaneously.

17.3 Results and Discussion

17.3.1 Worksheet About Human Blood

Traditionally, life sciences students rarely experience mathematics within the context of their majors. Indeed, anecdotal evidence suggest that students, who perceive themselves as mathematically weak, gravitate toward biology because they consider biology to be relatively mathematics free. These attitudes persist, despite mathematics course requirements for most life sciences students, because failure to integrate mathematics in meaningful ways for life sciences majors, contributes to the perception that mathematics is irrelevant outside itself (Zan, Brown, Evans, & Hannula, 2006). Thus, in the minds of many biology students, mathematics and biology remain in two distinct, separate compartments (Hester et al., 2014).

If students, who opt for biology because they envision a science career, but "don't like maths," they are pre-selecting to ignore mathematics. Adding a mathematical subject, or two, taught by mathematics experts to the university biology curriculum is very unlikely to bring about any positive changes, because such an approach would not help students discover the interdisciplinary connections by themselves (Ortiz, 2006; Šorgo, 2010). Increasing the quantitative thinking that teacher trainees encounter in their biology courses seems to be a more effective way to challenge this perception, and encourage the transfer of mathematical skills, compared to increasing traditional mathematics requirements for biology students alone (Hester et al., 2014). Therefore, some overlap ought to be found at the content level, and in teacher education. Mathematical-biology or biological-mathematics content knowledge for all teachers should be established (Šorgo, 2010).

An appropriate model, for re-working, turns out to be the model of pedagogical content knowledge proposed by Shulman (1986, 1987), upgraded to technological pedagogical content knowledge by Mishra and Koehler (2006), and used in cell biology by Usak (2009). The most important part of the model to be considered, is its central section referred to as Pedagogical Biological Mathematical Content Knowledge. In order to succeed in inter-linking these disciplines, it is inevitable to reform the general approach to the professional development of prospective biology teachers in terms of their pedagogical mathematical–biological content knowledge (Šorgo, 2010).

Therefore, we designed interdisciplinary material in which we targetted quantitative skills, that we identified as supporting the biological concepts taught in the course, and being generally useful to teacher trainees throughout their progress in advanced biology courses, and further teacher career. Table 17.1 categorizes each item by the biological concept and required mathematical competence.

Identifying biological learning outcomes of the material was relatively straightforward. Establishing quantitative learning outcomes of the worksheet required more consideration. When choosing an appropriate set of mathematical skills to be included in the material, we focused on the skills that students are expected to develop, by fulfilling their mathematical pre-requisites for the course, and integrate naturally

Table 17.1 Worksheet about human blood

<table>
<tr><th>Task</th><th>Biological content</th><th>Mathematical content</th><th>Task examples</th></tr>
<tr><td>1</td><td>Size of erythrocytes</td><td>Unit Conversion</td><td>Typical human erythrocyte has a biconcave shape with disc diameter approximately 7.5 μm and a thickness at the thickest point 2.1 μm and a minimum thickness in the centre 1 μm, being much smaller than most of other human cells
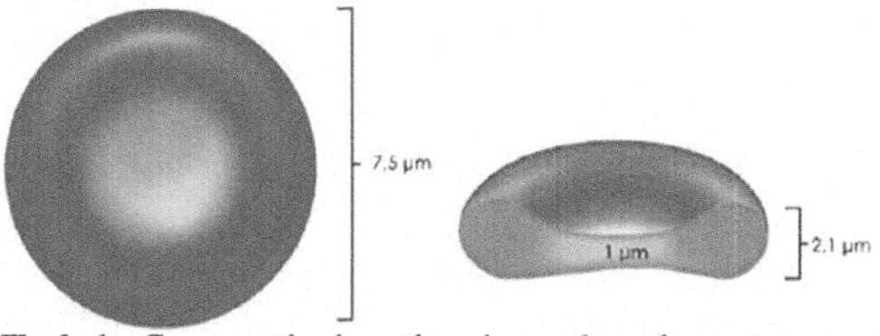

Task 1: Convert the length values of erythrocyte from micrometres to the most commonly used units of length, millimetres and metres</td></tr>
<tr><td>2</td><td>Volume of erythrocytes</td><td>Unit Conversion</td><td>Scientists use special units for measuring size of very small particles of human blood. For example: Volume of an erythrocyte is measured in femtoliters (fl)
Task 2: Volume of an erythrocyte is around 90 fl. Convert the volume of erythrocyte from fl to ml (use SI units table)</td></tr>
<tr><td>3</td><td>Blood volume of erythrocytes</td><td>Unit conversion</td><td>Number of erythrocytes is calculated from blood sample. Laboratory technicians count how many erythrocytes are in one mm^3 of blood or in one litre of blood. Table shows optimal amount of erythrocytes in 1 mm^3 of blood
<table><tr><th></th><th>Number of erythrocytes in 1 mm^3</th><th>Number of erythrocytes in 1 litre</th></tr><tr><td>Men</td><td>4.2×10^6–5.9×10^6</td><td></td></tr><tr><td>Women</td><td>3.8×10^6–5.2×10^6</td><td></td></tr></table>Task 3: Calculate the values of erythrocytes in 1 litre of blood, and fill in the table</td></tr>
</table>

(continued)

Table 17.1 (continued)

Task	Biological content	Mathematical content	Task examples
4	Geometric properties of erythrocytes	Percentages	Erythrocyte has biconcave shape and an average volume of about 90 fl with a surface of about 136 square micrometres **Task 4:** *Calculate as a percentage, how much the surface area of the biconcave shape of erythrocytes is bigger than hypothetical spherical cells having the same volume. Why is it important to consider the surface area of erythrocytes?*
5	Erythrocyte deformability	Percentages	The erythrocyte is a cell and its surface is called the cell membrane. Imagine that the biconcave shape of the erythrocyte changes into a sphere, while area of the cell membrane remains the same. The volume of such sphere would increase to 150 fl **Task 5:** *Calculate in percentage, how much the volume of erythrocyte increases if its shape changed to a sphere*
6	Haemoglobin in erythrocytes	Percentages	All haemoglobin molecules in one erythrocyte weigh together 32 ± 2 pg, which is around 34% of the erythrocyte weight **Task 6:** *What is the weight of an erythrocyte, if (32 ± 2) pg is 34% of its weight?*
7	Volume of haemoglobin in human body	Rule of three	Total number of erythrocytes in adult human body is usually between 2.10^{13} and 3.10^{13} and one erythrocyte contains around 265×10^{12} haemoglobin molecules. Haemoglobin is not contained in any other cell in a human body **Task 7:** *How many haemoglobin molecules are there in a human body?*
8	The erythrocyte life cycle	Rule of three	When erythrocyte gets old, it dies. In one minute around one million of erythrocytes die in human body. But do not worry, new cells are produced continuously and body keeps the number of erythrocytes within the range **Task 8:** *If erythrocytes did not regenerate, in what time would all erythrocytes in body die?*
9	Volume of iron in erythrocytes	Rule of three	Erythrocyte attaches to oxygen thanks to iron which is a constituent of haemoglobin. One gram of haemoglobin contains 3.34 mg of iron **Task 9:** *Calculate how many grams of iron are there in all erythrocytes in adult human body*

(continued)

Table 17.1 (continued)

Task	Biological content	Mathematical content	Task examples
10	*ABO* blood-group *genetics*	Combinatorics	**Task 10:** *Parents have four children. Is it possible that each child has different ABO blood type? Explain*
11	RH blood-group genetics	Combinatorics	**Task 11:** *Is it possible that Rh-positive parents have Rh-negative child? Explain*
12	Blood-group ABO genotyping in paternity testing	Combinatorics	**Task 12:** *A Mother with blood type B has a child with blood type A. Write down all possibilities of blood type which do not eliminate a man from paternity test*

into biology materials. Our motivation, for these two criteria, was that we wished to build quantitative learning outcomes that would be accessible to the biology teacher trainees, and that would support the biological learning outcomes, instead of being distracting and artificial.

Standard biology syllabi in Slovakia seldom make connections between biology and mathematics, and even those rare inter-links are just elementary. According to the biology curriculum, a successful biology teacher only needs to know how to calculate probabilities in Mendelian genetics (National Institute for Education, 2009). However, we identified other basic quantitative skills, namely unit conversion, calculation of percentages, and the rule of three. Once we had identified these skills, we discussed them with other experienced educators at CPU and project MaT^2SMc partner universities from abroad, interested in the integration of mathematics and biology. The university educators affirmed that they also consider it reasonable to expect teacher trainees to apply the skills we had identified. In addition, the educators expressed their disappointment caused by the teacher trainees' inability to do so in the contexts of their biology courses.

Consequently, the final version of the worksheet about human blood consisted of 12 items requiring four mathematical methods—unit conversion, calculation of percentages, the rule of three, and combinatorics (each of the methods required in three worksheet items)—implemented in biological context with respect to the secondary school curriculum. Our approach required no fundamental changes to the curriculum, and none of the pedagogical strategies used are revolutionary. We incorporated the application of 11 quantitative skills in the existing curriculum, and based the course design on recognized effective teaching practices eliciting active learning.

17.3.2 Mathematical Competence Assessment

The worksheet related to human blood consisted of items testing students' understanding of biology concepts, and their ability to apply mathematical skills in biological contexts. By means of content analysis of the teacher trainees' performance in each of the 12 items, their current mathematical competencies (unit conversion, calculation of percentages, the rule of three, and combinatorics) were evaluated (Table 17.2).

Apparently, the teacher trainees' scores vary with respect to the mathematical skills required by the tasks. In order to verify the significance of the differences, several statistical tests were conducted. First, since the assumption of normal distribution was not fulfilled, the non-parametric Friedman test was carried out. The calculated *p*-value was $p < 0.05$, thus the null hypothesis H_0*: The differences between the average scores with respect to the mathematical skills required by the tasks are not significant* was rejected at the 0.05 significance level. In other words, the differences between the biology teacher trainees' scores with respect to the mathematical skills required by the tasks are statistically significant (Table 17.3; Fig. 17.4).

Secondly, the Neményi method, of multiple comparisons, was performed to find where the differences actually occur (Table 17.4), i.e. in which pairs of the four mathematical skills.

The items with the highest incorrect answer rate (Table 17.3) were focussed on the application of the rule of three. The students' success rates were different significantly in these tasks than in tasks focussed on unit conversion. The students' success rates in tasks focussed on combinatorics differ significantly from tasks focussed on unit conversion, calculation of percentages, and the rule of three.

Table 17.2 Students' average scores in the mathematical tasks

Task	Task score[a]	Skill score
1	1.59	2.490
2	2.94	
3	2.94	
4	1.34	2.521
5	3.94	
6	2.28	
7	2.06	2.96
8	2.97	
9	3.69	
10	1.25	1.66
11	1.72	
12	2.03	

[a]1 = excellent; 2 = very good; 3 = good; 4 = fair and 5 = poor

Based on our own experience, and informal evidence from other biology lecturers, however, we questioned whether students were truly proficient at applying the required set of mathematical skills. By looking at teacher trainees' performance in each of 9 items focussed on unit conversion, calculation of percentages, and the rule of three (average rating 2.49–2.91), we conclude that their performance supported the assumption that students were unable to apply quantitative skills in the context of biology at the required level. This finding is consistent with the results of other researchers as it is widely recognized that students have difficulties when expected to transfer, spontaneously, skills to novel contexts (NRC, 2000), and that students in other fields are not able to transfer even relatively simple mathematical skills to new contexts (Britton, 2006). This result also supports earlier observations by Hester et al. (2014) that students perform significantly lower in tasks requiring application of mathematical skills to biological problems, than in context-free use of the same mathematical skills.

Most mathematics courses do not make a connection between mathematical concepts and applications to other fields of science (Robeva & Laubenbacher, 2009). Our observations underscore the need for interaction and collaboration to provide new alternatives to traditional mathematics and biology courses. It has been shown that a

Table 17.3 Friedman test results

	Rank sums	Mean ranks	Average score	Average score SD
Unit conversion	157.00	2.45	2.49	1.43
Percentages	171.00	2.67	2.52	0.87
Rule of three	200.00	3.13	2.91	1.16
Combinatorics	112.00	1.75	1.67	0.66

Test criterion value Q = 39.42; $p < 0.05$

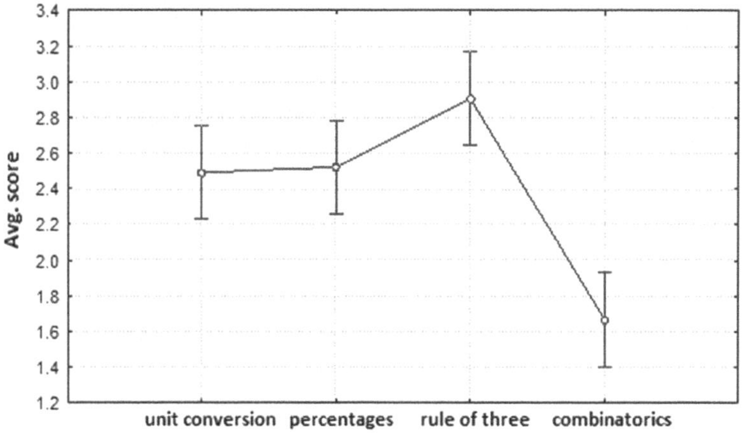

Fig. 17.4 Teacher trainees' average scores with respect to the mathematical skill

Table 17.4 Neményi test results

Unit conversion–Percentages	14.00
Unit conversion–Rule of three	**43.00**[a]
Unit conversion–Combinatorics	**45.00**[a]
Percentages–Rule of three	29.00
Percentages–Combinatorics	**59.00**[a]
Rule of three–Combinatorics	**88.00**[a]

[a]At a significance level of 0.05 the critical value is 37.49. Values exceeding the critical value indicate significant differences occurring in the compared pair(s) of mathematical skills required by the tasks

course integrating mathematics and biology concepts does not hinder student learning in either of the two content areas, but rather, enhances interdisciplinary knowledge (Madlung et al., 2011). Thus, we suggest that lecturers in mathematics and biology courses, convene to discuss these issues together. Such collaboration could lead to new mathematics courses, or supplementary instruction, involving biological examples in calculus and statistics courses. This can be done by sharing research data and, or, discussing interdisciplinary papers dealing with computational biology, and planning effective ways to present this information in their classes (Robeva, Davies, Hodge, & Enyedi, 2010; Watkins, 2010). In addition, lecturers can consider using, or modifying, the activities and materials developed within international projects, such as Compass, DynaMat, Primas, etc. (Čeretková, Jakab, & Naštická, 2015; Melušová et al., 2012; Pavlovičová, Rumanová, & Švecová, 2013; Sandanusová, 2006; Ulovec et al., 2009). The fact that students take mathematics requirements at different points in the curriculum, and come to biology courses with very diverse quantitative skills, is yet another challenge. To rectify this, we recommend that faculty members, in biology departments, re-evaluate the arrangement of the undergraduate curriculum, so that students take calculus and basic statistics during their first year of bachelor study. In this manner, they can move on to higher-level courses, such as genetics, microbiology, botany, zoology, and anthropology, with basic concepts that would allow them to apply mathematical tools in processing data obtained in laboratories, the interpretation of graphics, and improve their understanding of modern scientific theory. In accordance with our previous assumptions the students' success rates in tasks focussed on combinatorics were significantly different from tasks focused on unit conversion, calculation of percentages and the rule of three. Students achieved significantly better scores in these tasks focussed on combinatorics. Similarly, Colon-Berlingeri and Burrowes (2011) observed the best mathematics knowledge performance of biology students by looking at their performance on items measured basic probability, probability applied to genetics, and data interpretation relevant to the topics of Mendelian genetics, population genetics, and quantitative genetics. Authors (Colon-Berlingeri & Burrowes, 2011) suggest that the demonstrated mathematical knowledge in Mendelian concept may result from its interdisciplinary study within general biology courses. Our results, although limited by the small size of the sample studied, are in accordance with this suggestion.

Table 17.5 Results of the questionnaire survey

Statement		Likert 5-point scale[a]				
		1	2	3	4	5
The worksheet content and activities are understandable	Total	0	7	32	22	3
	%	0.00	10.94	50.00	34.30	4.69
The worksheet content and activities are interesting	Total	1	7	26	21	9
	%	1.56	10.94	40.63	32.81	14.06
The worksheet content an activities meet the curricular requirements	Total	0	5	26	30	3
	%	0.00	7.81	40.63	46.88	4.69
The worksheet meets teachers' practical requirements for teaching of the content	Total	1	7	21	24	11
	%	1.56	10.94	32.81	37.50	17.19
The worksheet could be used properly for collaboration of mathematics and biology teachers	Total	2	2	17	27	16
	%	3.13	3.13	25.00	42.19	25.00

n = 64; [a]1—strongly disagree; 2—disagree; 3—neutral; 4—agree; 5—strongly agree

17.3.3 Feedback Obtained in Questionnaire Survey

In order to receive authentic feedback from the biology teacher trainees, a questionnaire survey was carried out. The questionnaire consisted of 5 statements. The teacher trainees were asked to express the level of their agreement with the statements by means of 5-point Likert scale. The distribution of responses are summarized in Table 17.5.

The worksheet content, and activity understandability, were rated relatively low, with 50% of the respondents being neutral, and consequently, only 32.81% of students slightly agreed that worksheet content and activities were interesting. Also, nearly half of the teacher trainees agreed with the statement that the worksheet meets the curricular requirements (46.88%), but a slightly lower percentage (37.50%) agreed that teachers' practical requirements were met. The biology and mathematics collaboration, of content and activity, were rated very high, with slightly more than two-thirds agreeing or strongly agreeing (67.19%).

We had high expectations at the beginning of the interdisciplinary worksheet application. However, we consider the students' ratings of the worksheet as relatively low, which indicates a need for adjustment to the students' level of mathematical education at university. The survey results suggest that students can resist learning more mathematics than required for mathematics courses, and tend to avoid the biology lecturers who incorporate mathematics into their curricula. We believe that if an intensive and continual effort was developed among the departments, such that the true interdisciplinary links between mathematics and biology are highlighted in all courses across the curriculum, the students' attitudes could gradually ameliorate.

Student comments from the final worksheet survey were constructive. Representative comments are listed below:

- Add more maths details: maths formula where it is required, table, more illustrations, pictures, key words, blood illnesses.
- Too much mathematics, too much boring mathematics, too much difficult mathematics.
- Useful new knowledge about very small SI units (femtolitre).
- Nice example of interdisciplinary knowledge, useful for active thinking and deep knowledge and understanding.

Finally, we conclude that the analysis of feedback received from students provides the university educators with the basis for adjustments of the worksheet content and for improvements of the university educators' teaching strategies in interdisciplinary contexts.

17.4 Conclusion

Biology and mathematics teachers have much in common, in that they both seek to educate pupils to understand the wonder of the world around us. They both use the tools of logical thought, analysis, experimentation, and hypothesising to explore. But, too often, they see mathematics and biology as two distinct disciplines, separate from each other in objectives, methods, and classrooms. Also, all the pupils see mathematics and biology as completely different subjects, and often fail to apply lessons learned from one subject to another.

Therefore this study gives insights into continuing research regarding the assessment of the quality of biology teacher education in Slovakia, and development of educational instruments that would allow the acquisition of specific mathematical competencies necessary for future biology teachers in terms of an interdisciplinary approach. This study endeavours to bring these two allied subjects closer together. It attempts to promote, both biological and mathematical understanding, by ensuring both mathematics and biology teachers improve pupils understanding by working closely together.

Based on the findings of this study, we propose to offer an optional course to biology teacher trainees, that would cover several mathematical topics, and their applications in biology. The suggested mathematical, and related biological content, is shown in Table 17.6. We suggest that the lecturer of the course introduces the mathematical content and exemplifies its use in biology. In whole-class discussion, the teacher trainees should practise their mathematical skills while solving interdisciplinary tasks, assisted by the lecturer. As a follow-up, in pairs or small groups the teacher trainees should design similar interdisciplinary tasks and fit the tasks to the current primary and secondary school curriculum.

We believe that our work can serve as an example and inspiration for lecturers in different fields of biology in providing learning opportunities to bridge the gap between mathematics and biology, identifying specific areas that need to be improved, and devising alternative ways to address the gap.

Table 17.6 Proposal of interdisciplinary mathematics course for biology teacher trainees

Mathematical content	Biological content
Percentages	General knowledge applicable in most of the other areas
Unit conversion	
The rule of three	
2D and 3D shapes—symmetries	Plant and animal symmetries
2D and 3D shapes—surface areas and volumes	Leaf surface, Blood volume
Number sequences (Fibonacci numbers and the golden ratio)	Leaf arrangement, Number of petals in flowers, Shell spirals
Combinatorics	Mendelian inheritance
Ratios (magnification/reduction)	Work with microscope
Probability and elementary statistics	Experimental work, Population curve, Gaussian model, Mendelian inheritance
Exponential and logarithmic relations	Acidity of body fluids, Population growth and decay

Let us finish with an important statement: Mathematics and Biology teachers do a good, and often an outstanding, job in teaching young people the basic knowledge of their respective fields! It is not the intent of this study to criticize what they do or how they do it.

However, we observed that there is hardly any collaboration, or consultancy, between mathematics and biology teachers. Mathematics teachers often use a biology context in tasks, and biology teachers often use mathematics, mathematics used with little regard towards learning. The systematic posing of questions, organizing experiments or observations, and completing, analysing, and interpreting data, are everyday challenges for biology teachers. Mathematics, on the other hand, offers a universal tool for description and evaluation, as well as for modelling, real life situations. It is indisputable that every biology teacher should possess the basic mathematical skills that are necessary for his, or her, subject, and vice versa, every mathematics teacher should perceive pieces of knowledge and information, from other sciences, as an apt motivation for stressing the usefulness of mathematics.

Therefore, we developed materials that are useful for both mathematics and biology teachers, materials that can be used in teacher training, making biology and mathematics teacher trainees aware that working together, now and in their later careers, can improve their experience and the learning of their future students. These materials are designed to increase competencies in both subjects at the same time, allowing for interdisciplinary learning and for collaboration between biology and mathematics teachers, ranging from common lesson planning to team teaching.

Thanks to these materials, students and pupils are naturally involved in the studied content, they become more open in communication, and they start to apply their biology and mathematical knowledge in a more relaxed and natural way. It is therefore right to assume that the materials are likely to help both the prospective, and in-service, mathematics and biology teachers to present mathematics in a realistic biology context. We believe that study of our suggested topics, and solving the related problems, will be an interesting experience for all target groups. We hope that with this study we will encourage teachers to actively seek collaboration, regardless of whether they are a biology or a mathematics teacher.

Acknowledgements This work was supported by the Slovak Research and Development Agency under the contract No. APVV-14-0446, and by the LLP/Comenius MaT^{2}SMc project.

References

Anděl, J. (2003). *Statistical methods* (2nd ed.). Praha: Matfyzpress.

Bialek, W., & Botstein, D. (2004). Introductory science and mathematics education for 21st Century biologists. *Science, 303,* 788–790.

Brent, R. (2004). Points of view: The interface of mathematics and biology: Intuition and innumeracy. *Cell Biology Education, 3,* 88–90.

Britton, S. (2006). Are students able to transfer mathematical knowledge? In *Proceedings of the 2nd International Conference on the Teaching of Mathematics*. New York: Wiley. http://users.math.uoc.gr/~ictm2/Proceedings/pap267.pdf. Accessed December 12, 2016.

Čeretková, S., Jakab, I., & Naštická, Z. (2015). PRIMAS multiplier concept implementation in Slovakian context. In K. Maaß, B. Barzel, G. Törner, D. Wernisch, E. Schäfer, & K. Reitz-Koncebovski (Eds.), *Educating the educators: International approaches to scaling-up professional development in mathematics and science education* (pp. 56–65). Munster: WTM.

Choi, B., & Pak A. (2006). Multidisciplinarity, interdisciplinarity, and transdisciplinarity in health research, services, education and policy: 3. Discipline, inter-discipline distance, and selection of discipline. *Clinical & Investigative Medicine, 29*(6), 351–364.

Colon-Berlingeri, M., & Burrowes, P. A. (2011). Teaching biology through statistics: Application of statistical methods in genetics and zoology courses. *CBE-Life Sciences Education, 10*, 259–267. https://doi.org/10.1187/cbe.10-11-0137.

Gross, L., Brent, R., & Hoy, R. (2004). The interface of mathematics and biology. *Cell Biology Education, 3,* 85–92.

Hester, S., Buxner, S., Elfring, L., & Nagy, L. (2014). Integrating quantitative thinking into an introductory biology course improves students' mathematical reasoning in biological contexts. *CBE-Life Sciences Education, 13*(1), 54–64.

Hodgson, T., Keck, R., Patterson, R., & Maki, D. (2005). Mathematics in the biology classroom. A model of interdisciplinary education. *Journal of College Science Teaching, 34,* 46–50.

Jungck, J. (2005). Challenges, connection, complexities: Education for collaboration. In L. A. Steen (Ed.), *Math and Bio 2010: Linking undergraduate disciplines* (pp. 1–12). Washington, D.C.: The Mathematical Association of America.

Jungck, J., Gaff, H., Fagen, A., & Labov, J. (2010). "Beyond BIO2010: Celebration and Opportunities" at the intersection of mathematics and biology. *CBE-Life Sciences Education, 9*, 143–147.

Kalin, J., & Zuljan, M. (2007). Teacher perceptions of the goals of effective school reform and their own role in it. *Educational Studies, 33,* 163–175.

Madlung, A., Bremer, M., Himelblau, E., & Tullis, A. (2011). A study assessing the potential of negative effects in interdisciplinary math-biology instruction. *CBE-Life Sciences Education, 10,* 43–54.

May, R. (2004). Uses and abuses of mathematics in biology. *Science, 303,* 790–793.

Melušová, J., Šunderlík, J., & Čeretková, S. (2012). Interdisciplinary teaching of mathematics and science in Slovakia—pilot study. *Acta Mathematica, 515,* 95–100.

Mishra, P., & Koehler, M. (2006). Technological pedagogical content knowledge: A framework for teacher knowledge. *Teachers College Record, 108,* 1017–1054.

National Institute for Education. (2009). *State educational program—Biology (ISCED3).* Bratislava: Statpedu.

National Research Council. (2000). *Learning and transfer. How people learn: Brain, mind, experience, and school* (Expanded ed.). Washington, D.C.: National Academies Press.

National Research Council. (2003). *BIO 2010: Transforming undergraduate education for future research biologists.* Washington, D.C.: National Academies Press.

National Research Council. (2008). *The role of theory in advancing 21st Century biology: Catalyzing transformative research.* Washington, D.C.: National Academies Press.

Ortiz, M. (2006). Numbers, neurons & tides, Oh my! Mathematics, the forgotten tool in biology. *The American Biology Teacher, 68,* 458–462.

Pavlovičová, G., Rumanová, L., & Švecová, V. (2013). Geometry in gardens and parks. In J. Novotná & N. Mousoulides (Eds.), *Bringing mathematics to earth* (pp. 169–181). Nicosia: Demetra Ltd.

Pevzner, P., & Shamir, R. (2009). Computing has changed biology-biology education must catch up. *Science, 325,* 541–542.

Pursell, D. (2009). Enhancing interdisciplinary, mathematics, and physical science in an undergraduate life science program through physical chemistry. *CBE-Life Sciences Education, 8,* 15–28.

Robeva, R., Davies, R., Hodge, T., & Enyedi, E. (2010). Mathematical biology modules based on modern molecular biology and modern discrete mathematics. *CBE-Life Sciences Education, 9,* 227–240.

Robeva, R., & Laubenbacher, R. (2009). Mathematical biology education: Beyond calculus. *Science, 325,* 542–543.

Sandanusová, A. (2006). Interdisciplinary relations and their application in creating the worksheets. In *XXIV. International colloquium on the acquisition process management.* Brno: University of Defence.

Shulman, L. (1986). Those who understand: Knowledge growth in teaching. *Educational Research, 15,* 4–14.

Shulman, L. (1987). Knowledge and teaching: Foundations of the new reform. *The Harvard Educational Review, 57,* 1–22.

Šorgo, A. (2010). Connecting biology and mathematics: First prepare the teachers. *CBE-Life Sciences Education, 9*(3), 196–200.

Steitz, J. (2003). Bio 2010: New challenges for biology educators. *Cell Biology Education, 2,* 87–91.

Stock, P., & Burton, J. (2011). Defining terms for integrated (Multi-Inter-Trans-Disciplinary) sustainability research. *Sustainability, 3,* 1090–1113.

Tra, Y., & Evans, I. (2010). Enhancing interdisciplinary mathematics and biology education: A microarray data analysis course bridging these disciplines. *CBE-Life Sciences Education, 9*(3), 217–226.

Ulovec, A., Čeretková, S., Dockerty, A., Molnár, J., & Spagnolo, F. (2009). *Motivating and exciting methods in mathematics and science—Case studies.* Olomouc: VUP.

Usak, M. (2009). Preservice science and technology teachers' pedagogical content knowledge on cell topics. *Educational Sciences: Theory & Practice, 9,* 2033–2046.

Viteckova, M., Prochazka, M., Gadusova, Z., & Stranovska, E. (2016). Identifying novice teacher's needs—The basis for novices' targeted support. In: L. Gómez Chova, A. López Martínez, I. Candel Torres (Eds.), *Proceedings of ICERI2016 Conference* (pp. 7731–7738). Seville: IATED.

Watkins, J. (2010). On a calculus-based statistics course for life science students. *CBE-Life Sciences Education, 9,* 298–310.

Williams, J., Roth, W.-M., Swanson, D., Doig, B., Groves, S., Omuvwie, M., et al. (2016). *Interdisciplinary mathematics education: State of the art.* Cham: Springer. https://doi.org/10.1007/978-3-319-42267-1.

Zan, R., Brown, L., Evans, J., & Hannula, M. (2006). Affect in mathematics education: An introduction. *Educational Studies in Mathematics, 63,* 113–121.

Chapter 18
Creating Academic Teacher Scholars in STEM Education by Preparing Preservice Teachers as Researchers

Jennifer Wilhelm and Molly H. Fisher

Abstract We describe what Research Experiences for Undergraduates (REU) Fellows reported regarding their experiences within a research-intensive programme in STEM Education. Our study employed a mixed methods approach. Quantitative data included a survey by Kardash (2000) to examine Fellows' research expectations, familiarity with literature, and ability to conduct statistical analyses pre and post the interdisciplinary STEM Education programme. We also analysed qualitative data from the Fellows' pre-, mid-, and post-evaluation interviews. We discovered significant growth in their confidence levels in studying, conducting, and analysing research ($p < 0.001$). At the end of their nine- month research experiences, Fellows stated they felt they had gained skills in coding and analysing data, conducting interviews, using technology, writing, and presenting, but most frequently noted their increase of interpersonal collaborations with other future STEM teacher researchers. This research is the first to examine the effectiveness of an academic year interdisciplinary STEM Education REU programme. REU programmes typically are offered two months (8 weeks) between Spring and Autumn semesters and within only one STEM content discipline located in schools of science and engineering, as opposed to education.

Keywords Mathematics educational research · Science educational research · Undergraduate research experiences · Pre-service teacher education · Pre-service teacher professional development

18.1 Introduction

Our novel Research Experiences for Undergraduates (REU) project exposed 23 preservice K-12 science, mathematics, elementary, and special education teachers to

J. Wilhelm · M. H. Fisher (✉)
University of Kentucky, Lexington, USA
e-mail: molly.fisher@uky.edu

J. Wilhelm
e-mail: jennifer.wilhelm@uky.edu

B. Doig et al. (eds.), *Interdisciplinary Mathematics Education*,
ICME-13 Monographs, https://doi.org/10.1007/978-3-030-11066-6_18

timely problems involving STEM teaching and learning through original research conducted with eight successful mathematics and science Education faculty mentors affiliated with the University of Kentucky's STEM Education Department in the United States of America (USA). This project expands the research base concerning undergraduate student research across the disciplines, and serves to inform international efforts towards developing future K-12 STEM educators able to implement research-confirmed pedagogical practices, conduct and respond to authentic research concerning their practice, and serve as agents of change regarding local schooling realities. Our REU project's main objective for pre-service K-12 teachers was to instill an appreciation of the interplay between society, education, and STEM research, that can lead towards improving K-12 students' levels of STEM knowledge and application.

This study aimed to answer the following questions: (1) To what extent can undergraduate STEM Fellows' knowledge and confidence in STEM Education research change through participation in an intensive nine-month research programme? And, (2) What research products, and career trajectories, will result from REU Fellows' participation in our interdisciplinary STEM Education programme?

We designed, implemented, and researched, the effectiveness of our current REU Site project at the University of Kentucky (UoK). Our project: (1) expanded research efforts into novel phenomena in K-12 STEM Education, (2) delivered quality STEM education via Fellows' and faculty mentors' research, and outreach activities, embedded within authentic interdisciplinary STEM formal and informal environments, and (3) created a cadre of educators poised to teach mathematics and science effectively, appreciate the interdisciplinary nature of STEM phenomena, research their practice, and serve as leaders in STEM education at the elementary, middle, and secondary school levels.

18.2 Background and Literature

STEM is no longer simply an acronym for Science, Technology, Engineering, and Mathematics, but also a unification of the disciplines that, as an entity, is much greater than the sum of its parts. The field of STEM educational research is ripe with potential for exploration of issues across these disciplines, in high synergy with one another, in terms of the importance of student learning and citizen participation. The barriers towards these things is now of great concern to many, as the USA continues to fall in its international economic competitiveness (Committee on Science, Engineering, and Public Policy, 2006). The under-preparation of the USA's workforce, with respect to STEM knowledge and skills, is a very real problem to remedy, especially when one considers the performance of USA school children across the STEM disciplines, compared to their international peers. The 2006 PISA, which examined 15-year-olds from 57 countries, showed average performance of USA students in terms of scientific concept recall (level 1) and significantly below average performance regarding understanding and application of mathematical and scientific concepts (levels 2 and 3). In contrast, Canada (a neighbouring country) ranked in the top 5 in all three categories (OECD, 2007). The 2012 PISA results showed a similarly bleak picture. Students in the USA performed below average in mathematics in 2012, ranking 36 out of 65 countries and economies participating in PISA. USA students performed near average in both reading and science, ranking 23rd in reading and 28th in science (OECD, 2014). For the 2015 PISA results, USA reading and science scores remained relatively the same as the 2012 scores; however, the USA dropped 11-points in the average mathematics score causing the USA to rank 38th out of 70 participating countries and economies (OECD, 2016).

Equally depressing are postsecondary students' performance with respect to mathematics and science understanding and application (Committee on Science Engineering and Public Policy, 2006; US Department of Education, 2006; US Office of Science and Technology Policy, 2006). Kentucky ranked 49th in the USA in 2003 for bachelor's degrees in science or engineering conferred (Kentucky STEM Task Force, 2007). Our project's ultimate goal is an improvement of K-12 students' levels of STEM knowledge and application, especially for students from historically disadvantaged groups, who enter universities with an especially negative stance towards STEM and their ability to participate in STEM (Clewell, Anderson, & Thorpe, 1992; Gatta & Trigg, 2001). The creation of a larger and better-prepared K-12 mathematics and science teacher population is our main strategy towards meeting our goal.

Our REU site, the UoK STEM Education Department, is uniquely positioned at a research university. Our REU research activities are united by their applicability to real-world pedagogical problems in actual STEM learning environments. Poised to train the next generation of teacher researchers are the eight full-time mathematics and science education faculty researchers. Our faculty represents diverse, but overlapping, research agenda, linked by an interest in STEM education that focus on a variety of content areas (mathematics, physics, Earth/space, chemistry, and engineering) as well as all levels of K-20 education. Our collective work explores the multiple means of encouraging effective, deep STEM learning, emphasizing the interdisciplinary ways of understanding physical phenomena, and creating active research participation in STEM by K-20 learners through quantitative, qualitative, and mixed methodologies. With our faculty serving as their mentors, REU Fellows gained experience with current educational issues, understanding of unifying conceptual ideas, and methodological strategies underlying STEM educational research (including its applicability to real-life scenarios), and awareness of the links between society and schooling.

Our REU project redefines the focus of STEM teacher preparation towards the development of students' higher order thinking skills, and deep conceptual STEM understanding, based on research-confirmed best practices discovered by the preservice teachers themselves. Our basic pedagogical approach centres on the active participation of Fellows in authentic inquiry into local educational issues, with national and international implications. Undergraduate participants are engaged in activities that research has demonstrated to be most important with respect to undergraduate research, including reviewing pertinent scholarly literature, developing testable and timely hypotheses and associated empirical tests, gathering and evaluating related data, and communicating their research to others (Landrum & Nelsen, 2002; Lopatto, 2003; Sabatini, 1997). Previous research has shown that undergraduate research programmes can allow students to make a more educated decision regarding the pursuit of a graduate degree (Willis, Krueger, & Kendrick, 2013). Additionally, Kardash (2000), found that female interns in undergraduate research experiences showed less confidence in certain aspects of research than male interns. In the predominantly female field of education, this type of research can be crucial in order to gain more females in STEM fields as "mastery of skills predicts efficacy beliefs, which in turn predicts career aspirations" (Adedokun, Bessenbacher, Parker, Kirkham, & Burgess, 2013). More confidence in both the areas of STEM content and STEM educational research, for teachers of mathematics and science, should lead to a better prepared student population. Our REU programme is unique, since it extends throughout an academic year, as opposed to the typical 8-week programme that occurs between a Spring and Autumn semester. In addition to this, our programme emphasizes and creates interdisciplinary STEM experiences as Fellows learn, teach, and research STEM content.

18.3 Participants

During a three-year period, we recruited 23 highly committed undergraduate students as Fellows, who functioned as a cohort within the larger dynamic of our STEM educational research community. The Fellows represented a diverse group of undergraduate students studying elementary, middle, and secondary level mathematics and science education. In the first cohort, our student demographics were the most diverse with three African–American students (two of whom were "non-traditional" students in terms of age), five White students, one male (which is not a surprise for a programme housed in a College of Education in which the students are predominantly female), five secondary STEM Education majors (four mathematics and one science), two elementary education majors, and one middle level mathematics education major. The second cohort was the least diverse, with eight White female students, in which four were secondary mathematics education majors and four were elementary education majors. The third cohort consisted of one male, one Hispanic student, seven White students, three secondary mathematics education students, three elementary education students, and two special education students. The third cohort included one Fellow from the previous cohort who returned to the programme for a second iteration, thus, the reason for 23 total Fellows. The inclusion of Fellows from major programmes, that are not specifically a STEM content based area, allowed the programme to spread awareness of STEM educational research beyond just students in our department, and has the potential to have an impact on the Fellows' STEM content and pedagogy, while undergraduate students, and in their future classrooms.

18.4 Programme Description

Throughout the programme, Fellows worked in pairs. Each pair was mentored, by one to three faculty researchers (for researchers working on common projects) in the Department of STEM Education (one additional mentor was from an off-site institution). Fellows met with their mentors weekly in Autumn and Spring for 1–3 h to discuss the progress of their research projects, ask questions emerging from the research procedures, and determine the subsequent research tasks. When mentors, or mentees, could not attend the meeting in person, or on campus, virtual reality, or remote communication tools, were used in order to proceed with the meetings. Most of the Fellows viewed these meetings as great opportunities to develop understanding about their research projects, through interacting with their mentors, graduate students, and their peers. In addition to weekly meetings with their mentors, the Fellows participated in a weekly seminar with the other Fellows. Fellows were interviewed by the project evaluator at the beginning, middle, and end of their experience.

18.4.1 Individualized Research Projects

Fellows, in groups of two, worked on individualized research projects under the direction of their faculty mentor(s) conducting research of interest to the Fellows. Fellows met weekly with faculty mentor(s) during the Autumn and Spring semesters. Discussions and work between fellows and mentors were guided by a set of research activity guidelines developed by the Principal Investigator and Co-Principal Investigator to ensure that all projects were progressing efficiently, and that all Fellows' metacognition about STEM Education research was normalized, to some degree, across the cohort. Fellows worked with their mentors on conducting a review of literature, securing IRB approval, developing their hypotheses and empirical tests (Autumn), data collection (Autumn and Spring), and analysis and presentation (Autumn and Spring). Presentations of findings included students' preparation of research talks, posters, and papers, to be presented at their cohort seminars, and at both the Student Research Conference as well as one other research conference with their faculty mentors. Fellows and faculty mentors were also encouraged to submit for publication, at least one article to a practitioner, or research journal. Four projects were conducted each year (with two Fellows *per* project). Below is a sample of four student projects that illustrates the Fellows' research as well as the interdisciplinary nature of the STEM content that was analysed—including spatial visualization in Earth/Space science, computer programming, and engineering design within a robotics laboratory, mathematical experiences that affect STEM attitudes and beliefs, and the unique ways of attending, interpreting, and deciding within STEM classrooms:

- *Exploring variables that affect students' scientific and spatial understandings as they engage in Earth/Space science*: This research examined the differences between groups of middle school students' spatial and scientific reasoning from pre- to post-implementation of an Earth/Space unit. Using a quasi-experimental design, researchers explored how instructional methods and sex affected learning. Treatment teachers employed a project-based curriculum while control teachers implemented a traditional Earth and Space unit. A Lunar Phases Concept Inventory (LPCI; Lindell & Olsen, 2002), the Geometric Spatial Assessment (GSA; Wilhelm, Ganesh, Sherrod, & Ji, 2007), and the Purdue Spatial Visualization Test-Rotation (PSVT-Rot; Bodner & Guay, 1997) were used to assess learning. To gain a deeper understanding of students' spatial-scientific reasoning, lessons related to the Earth/Space unit were video-taped and students participated in video-recorded clinical interviews. Interviewees were selected purposefully, based on their highest and lowest scores on the LPCI from pre- to post-test, allowing researchers to examine high and low performing students' spatial development and scientific understandings of lunar phases by sex within, and between, control and experimental groups. This is the first known study that investigated the spatial skills of low and high performing students before, during, and after an Earth/Space unit. REU Fellows participated in data collection and analysis at various levels, including the classroom observations, test administration, clinical interviews, video-taping, transcribing, and analyzing, qualitative and quantitative data. Fellows were mentored on designing an interview protocol that examined how middle school students reason, negotiate the subject matter, and apply learned concepts to new situations. Fellows independently conducted the interviews, transcribed, and analysed results.

- *Informal learning environments in STEM education*: This project examined the effect of an informal STEM learning environment on middle grades students. The STEM informal learning environment, in this project, was designed to address the inequities of under-represented populations (females and students of colour) experience in formal learning settings, with an underlying goal to increase students' interest and motivation to learn interdisciplinary STEM content, and participate in STEM-related activities. Interdisciplinary learning experiences included robotics, which purposefully integrated computer programming and engineering design. Research Fellows coded and analysed written survey responses obtained from students after the conclusion of their informal experience. These data indicated an increase in students' interest in STEM, and an appreciation of the authentic STEM activities that permitted the students to feel like an engineer, scientist, and, or, mathematician.
- *Using the Mathematics Experiences and Conceptions Survey (MECS) to understand pre-service Elementary teacher's attitudes and beliefs towards STEM*: This work comprised analyses of data from the Mathematics Experiences and Conceptions Survey (MECS; Jong, Hodges, Royal, & Welder, 2015) designed to understand the evolution of pre-service elementary teachers' attitudes, beliefs, and dispositions towards mathematics and applied mathematics in teaching and learning. Two characteristics that make the surveys particularly powerful are that they: (1) are designed to measure conceptions longitudinally, and (2) collect extensive demographic information, along with items about the pre-service teachers' past experiences as learners of mathematics. The Fellows coded open- ended responses, and learned about descriptive statistics to examine demographic data collected

on the pre-service teachers. The qualitative and quantitative analyses were both completed using data analysis software that provided the Fellows with essential research skills to develop codes that informed patterns of themes based on frequencies. They found that pre-service teachers attributed STEM teaching success to the following three factors: differentiating instruction, creating engaging lessons with purposeful tasks, and creating a positive classroom community.

- *Professional noticing of STEM thinking*: Professional noticing is the study of investigating what teachers notice and attend to in classroom situations, how they interpret this information, and what they decide and prescribe as next steps in assisting students in moving forward in their STEM learning (Jacobs, Lamb, and Philipp, 2010). This project consists of a multi-institutional collaborative working towards refining pre-service teachers' practices in STEM education contexts. Specifically, this project involves the creation of focussed course materials (including technology-mediated resources) aimed at the development of professional noticing capacities, and study of pre-service teachers' growth in such dimensions. Fellows participating in this project collected, and analysed, data from other pre-service teachers participating in a professional noticing instructional module focusing on STEM professional noticing skills. This project was carried out multiple years in varying capacities with different Fellows participating each year. The data collected included analysis of written responses measuring pre-service teachers' professional noticing skills, scoring and analyses of content knowledge changes within the semester when the instructional module was taught, and interviewing former pre- service teachers that participated in the module and who are now classroom teachers. Fellows found that teachers retained their knowledge in professional noticing when teaching STEM content in their own classrooms, despite the fact they may not remember the exact terminology of the professional noticing construct. In terms of content knowledge, pre-service teachers were found to have increased in some aspects of their content knowledge, especially when trying to explain a child's thinking on certain problems.

18.4.2 Methods

We examined the following research questions: (1) To what extent can undergraduate STEM Fellows' knowledge and confidence in STEM educational research change through participation in an intensive nine-month research programme? (2) What research products, and career trajectories, will result from REU Fellows' participation in our interdisciplinary STEM education programme?

This research employed a mixed methods approach (Creswell & Plano Clark, 2007): (a) Quantitative data included administering a pre- and post-survey with a Likert scale (Kardash, 2000) concerning undergraduate research, to determine Fellows' research expectations, familiarity with research literature, and ability to conduct statistical analyses, and (b) Qualitative data was comprised of the Fellows' pre-, mid-, and post-evaluation interviews, as well as their final research projects, papers,

Table 18.1 Examples of Survey Questions (Kardash, 2000)

Type of question	Example questions
"To what extent can you… (24 Questions)	… write a research paper for publication in the field of education? … analyse data statistically in the area of research you are working on in this programme?
"To what extent do you feel this internship will help you… (23 Questions)	… identify a specific question for investigation based on the research in the field of education? … observe and collect data in the field of education?
Other (6 Questions)	I have the ability to have a successful career as a STEM educator I possess the motivation and persistence required for a career as a STEM educational researcher

and presentations. Questions on the Kardash (2000) survey included items having to do with Fellows' beliefs concerning how well they felt they understood concepts in the field of education, and in their area of research, and how skilled they felt about making use of scientific research literature in the field of education, and in their area of research (see Table 18.1). Qualitative interviews included questions regarding how well, and in what ways, the REU programme advanced Fellows' STEM Education research knowledge, and had an impact on their future teaching in the areas of mathematics and, or, science.

18.5 Results and Discussion

Using a two-tailed t-test to analyse the results of the quantitative survey, we discovered significant growth in the Fellows' confidence levels in studying, conducting, and analysing research. This growth was also echoed in their interviews with the programme evaluator. Further, our findings showed Fellows entered the programme with confidence in analysing literature, but still needed improvement in the areas of designing their own research projects, and performing statistical analysis of their research data. Despite their initial confidence, Fellows believed this programme would increase, greatly, their abilities to conduct research and write manuscripts for publication. At the end of their nine-month research experiences, Fellows stated they felt they had gained expertise in coding and analysing data, conducting clinical interviews, using technology, writing, and presenting, but, most frequently, they noted the increase of interpersonal skills and collaboration with peers. Table 18.2 shows the results of t-tests for each year of the programme, as well as a combined test of all cohorts. While some may quibble over the use of a t-test in lieu of a non-

Table 18.2 t-table with pre- and post-programme changes by cohort and combined[a]

	Pre	Post	Change	Significance
Year 1	4.11	4.48	0.37	$t(7) = -2.805, p = 0.026$
Year 2	4.10	4.36	0.26	$t(7) = -2.096, p = 0.074$
Year 3	4.08	4.47	0.39	$t(5) = -2.612, p = 0.048$
Combined	4.10	4.44	0.34	$t(21) = -4.488, p < 0.001$

[a]In year 3, two Fellows did not participate in the survey data (one was the Fellow that repeated the programme)

parametric test, de Winter and Dodou (2010) found no difference between a t-test and a non-parametric Mann-Whitney-Wilcoxon test when using five-point Likert data.

18.5.1 Fellows' Beliefs Concerning Stem Education Research

The REU programme clearly influenced students' beliefs about educational research. When they were asked about their views on educational research in pre-programme interviews, no Fellow mentioned the importance, or nature, of educational research: however, Fellows' view on educational research went much deeper by the middle and end of their REU experiences. During the mid-point interviews, the Fellows began to express the view that they realized educational research is highly important as educational research improves knowledge of instructional methods, and helps to make informed educational decisions. Excerpts highlight the points as below.

(Excerpt 8: mid-interview)

Interviewer: what is your current view on educational research? Fellow: That's real important. I think that education is something that is not seen as high as, for example, some type of science research, but actually I think it is as important, especially since there's a lot of reform with education, like what's best … so I think research really helps to decide what the best thing to do for education is and how to reform it. It's constantly changing.

(Excerpt 9: mid-interview)

I believe it's to make the whole process of education for both teachers and students a more productive experience … enhancing education on both sides. That's what I see educational research as … just making the whole institution of education better.

At the end of the year-long experience, the Fellows expressed that they realized (1) a great number of interdisciplinary topics can be studied in STEM education, (2) it is very important for teachers to keep up with the current educational studies to improve their teaching, and (3) it is possible for teachers to conduct educational research even though it is not easy. An example follows:

(Excerpt 10: post-interview)

I think it's really important. Like, as a nation we have so many different ideas as to what is the best way to teach students, and there are so many different learning styles that are prevalent now that weren't even thought of maybe 20 years ago. I think it's really interesting to keep finding out the ways students learn because school is … You know, the majority of their lives growing up they spend so much of their time there and then pursue higher education, and kids can spend up to 10 to 12 years in school. So it's really important to keep researching, and keep understanding how students learn, how we can teach better … it's really important. Sometimes it's harder because it's educational, so it's a lot of observing in classrooms, it's a lot of transcribing interviews, it's not like a scientific thing where you administer this and then you jot down the data, and the numbers will give you your answer. So it takes a lot more of an "analysational" approach, and it's harder and it's longer, but I think it's definitely really important."

18.5.2 Student Skills of Education Research

To assess the change in Fellows' skills of educational research, students were asked to rank their research skills and abilities in both pre- and post-surveys using a Likert scale ranging from "No skills and ability (1)" to "Great skills and ability (5)." During the final phase of the project, the same questions were given to a control group (non-participant students from another educational course) of similar size as our final REU cohort. The Fellow completing her second year was excluded from this analysis to maintain data consistency. Fellows' skills were assessed on the following aspects: (a) Thinking and working like a researcher, (b) Personal gains related to research work, (c) Research skills, and (d) Attitude and behaviours as a researcher.

Nine statements were used to assess students' gains within the "Thinking and Working like a Researcher" aspect and those statements were examined more closely. At the end of the year-long experience, all final cohort Fellows indicated that they possessed good skills, and, or, abilities with respect to all nine statements. Of these aspects, Fellows significantly gained in *(1) identifying limitations of research methods and designs, (2) understanding the theory and concepts guiding educational research,* and *(3) formulating a research question that could be answered with data*. Results indicated that they gained the least in *understanding how educational research projects connect to the broader teaching challenges in the field* and *problem-solving in general*. The limited gains were due, most likely, to Fellows' high confidence in these aspects at the programme's beginning. Compared to the control group students, Fellows possessed high level skills and abilities in all nine aspects as shown in Fig. 18.1.

18.5.3 Impact on Fellows' Future Teaching

Fellows' responses to the post-survey suggested the REU experience did not significantly change their career goals. Most of the Fellows still planned to work as a teacher at the K-12 level. However, the post-interview suggested the experience had helped students gain deep pedagogical knowledge, and strategies, that could be integrated into their future STEM teaching. The experience also had a strong influence

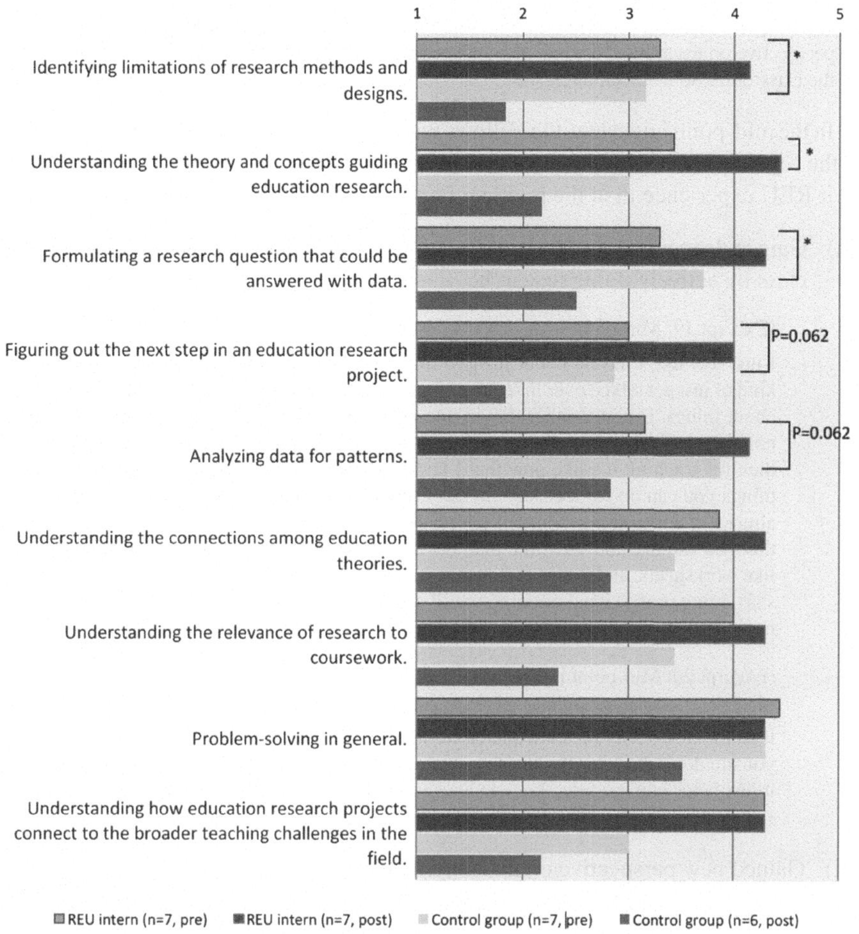

Fig. 18.1 Gains in thinking and working like a researcher ($^*p < 0.05$)

on Fellows' perspective on their teaching career, and motivated them to conduct their own research as a teacher. As two Fellows noted in the post-interview:

(Excerpt 13: post-interview)

Our specific research was about a type of pedagogy, and so I think that I am very … I'm pretty much an expert on that subject now after all of the things that I've read. So I think that I could probably use that in my classroom, and then again my experience will help me with my research – knowing which types of research I want to do, how to go about it, what type of interview I want to do, so just getting that information from the programme will be beneficial in the future because I probably want to do it myself.

(Excerpt 14: post-interview)

I don't plan on doing anything STEM necessarily. But it has given me, I guess, a better perspective on teaching and what all goes into that, especially with professional noticing in the classroom. So it's been interesting.

In the mid-point interview, the Fellows were asked to compare the REU experience to the regular education classes. All Fellows expressed that they gained more from their REU experience than the regular classes, specifically in the following ways:

(i) Gained deeper insight and understanding about the knowledge taught in regular class by actively doing research:

(Excerpt 19: Mid-point interview)

I just feel like I have a better insight, like I have more knowledge about things that kind of just get talked over in regular classes. Like I have more in depth understanding about things. Like the other day in one of my classes we talked about professional noticing, but it was brought up very … it's just like, almost like one day in class and then for teaching it's like now that I know all about it I feel like it's one of the best things you can do for a classroom. So I feel like that should be more of a thing. It's almost … it is a thing … in all the regular ed classes they talk about the strategies that you can use and stuff like that, but this is something that like … like strategies, like worksheets, and things to do in the classroom … but this is something that if you start doing professional noticing you'll start to be a better teacher more so than just having all the paperwork ready. You'll be able to teach well, teach better off.

(Excerpt 20: Mid-point interview)

It's just way more hands-on, you can actually see it yourself rather than just reading it from a textbook and someone else kind of telling you what the research is. When you are doing it yourself you can … you're actually in the depth of it, and you are … it just helps you get more knowledge and more personal experience with the subject … whatever it would be.

(ii) Gained new perspective on teaching methods:

(Excerpt 21: Mid-point interview)

I think it helps you see teaching in a different way. Like when you're in your normal classes, you see teaching from observing an actual teacher; and doing the research you see … you get to see so many studies done on teachers in their classrooms, and you get to see the background … like more technical side of teaching than just observing a classroom. So, like you learn about the research behind pedagogical strategies and things like that that can really benefit you more than just watching someone teach and not having like these terms and the research to backup if that's good or bad.

(iii) Gained understanding about practice and school atmosphere:

(Excerpt 22: Mid-point interview)

I feel like I've gained so much more understanding about how students learn and think, and then definitely this air of professionalism going into schools, working with teachers, communicating with them regularly … I feel like I'm sort of on their level as opposed to like just a student coming into watch their class. So it's definitely given me a lot of experience other than just that passively sitting in the back of the room, taking notes and stuff.

The overall productivity level of the Fellows in the project has been astonishing. To date, REU Fellows involved in this programme have participated in fifteen research presentations, and have co-authored nine manuscripts (that are printed or in press) in journals such as School Science and Mathematics and Science Scope, with many other manuscripts currently under review, or under revision, for a journal. The accolades of the Fellows do not end with scholarship in terms of presentations and publications. Four Fellows have been awarded summer research grants to continue their work, one was offered a summer position teaching for a prestigious summer programme at Duke University, three entered graduate programmes with fully funded research positions (one Master's programme and two doctoral programmes), one worked as a pre-service teacher summer research Fellow for Pearson, and another went on to participate in a summer REU experience at another research university after leaving her academic year REU programme at UoK.

One of the Fellows, from the first cohort, summed up her experience with the following:

> I definitely think about research differently from what I did before. Before it was like this intangible thing and too hard. But actually getting involved and doing it was a really great experience. It also was seeing the importance of doing the research and looking at the results and how those can be used in the classrooms and how that is beneficial to the education community in general.

Although, their ultimate career goal of becoming a mathematics or science teacher did not change, five of the Fellows did choose a trajectory that included additional research experiences in the area of STEM, and three of the Fellows went on to pursue graduate degrees prior to becoming classroom teachers.

18.6 Conclusion and Coda

As shown in both the quantitative and qualitative data, STEM Fellows' knowledge and confidence in STEM educational research increased. Research projects were numerous and included conference presentations and publications. Similar to Willis et al. (2013), our programme allowed the Fellows to make more educated decisions regarding pursuing graduate degrees and embarking on new research endeavours.

Our study added to the Kardash (2000) study in several ways. Kardash noted female REU interns tended to show less confidence in research skills than their male counterparts. Our research found females improving in their research skills in multiple areas, giving them the ability to be, not only more effective teachers, but also, more effective teacher researchers. Perhaps being female REU Fellows in a predominantly female field, such as education, might have helped with their confidence. Kardash claimed that one of the limitations of her study was the lack of a control group. During the final phase of our interdisciplinary STEM Education programme, a control group (non-participant students from another education course of similar size to our REU cohort) was added. Fellows' gains on thinking and working like a researcher, were higher than that of the control group especially in the areas of *(1) identifying limitations of research methods and designs, (2) understanding the theory and concepts guiding educational research,* and *(3) formulating a research question that could be answered with data.*

While this study does build upon the findings of Kardash (2000), it also has many differences. Most notably, the Kardash study focussed on a group of student interns majoring in science fields. These students spent research hours in a science laboratory conducting research in areas such as biology, biochemistry, chemistry, and physics. Fellows, in our present study, were participating in social sciences research with human subjects, which had STEM content embedded in the research projects. Additionally, Kardash used groups of Fellows from two very different undergraduate research experiences. Of the 57 interns studied, nineteen were from an academic year programme in which the students worked twelve hours per week and the other 38 Fellows were from a Summer (two month) programme in which the students worked 40 h per week in the science laboratories.

We recognize that there are limitations to our study as well. Mostly, the sample size of eight interns *per* year, is small, and could contribute to possible statistical errors. Also, with the 23 Fellows being assigned to eight different faculty mentors throughout the course of the three-year project, the Fellows were afforded a variety of research experiences. Most Fellows participated in qualitative research, but a few participated in quantitative and mixed methods research. While not entirely a limitation, the diversity in experiences meant that some Fellows did not get as much practice across the different types of research methodologies. On the other hand, this allowed for more fruitful and diverse discussions during the class meetings of the Fellows, as they discussed their progress in their research projects.

This REU project is original and unique, as it is one of the first where participating students are preparing to be K-12 teachers, as opposed to preparing for a career in a STEM field. In addition, this REU project was enacted during the academic year, whereas most REU opportunities occur during the two-month Summer break. Working with the Fellows during the academic year provided opportunities for Fellows and their mentors to conduct their research in formal and informal educational settings with human subjects. Our STEM Education faculty represent diverse, but overlapping research agenda, linked by an interest in interdisciplinary STEM education that includes motivation and career trajectories for STEM educators in K-12 and higher education, issues of racial and sex equity, how to increase participation and connection-making in K-12 mathematics and physical sciences, and implementation of technology, and engineering concepts, and tools that unite, and help teach, across STEM content areas. Our work explored, collectively, the multiple means of encouraging effective deep STEM learning, and active participation in STEM content, by K-20 learners through quantitative, qualitative, and mixed methodologies. With the STEM Education faculty serving as their mentors, REU Fellows gained experience with particular educational problems, broad understanding of unifying conceptual ideas, and methodological strategies, underlying STEM educational research (including its applicability to real-life scenarios), and awareness of the links between society and schooling. Future research with REU Fellows will include determining how the Fellows' research experiences influenced their teaching and classroom practices.

Our interdisciplinary STEM research programme can inform anyone that is working with undergraduate researchers, even those working in shorter time periods, and not working with human subjects as our projects have. This programme has now created 31 teacher researchers in four cohorts of Fellows. Through social media, and other measures, we have managed to stay in contact with the majority of the Fellows. Many of them have re-iterated the importance of our programme and how their knowledge has translated to their classrooms. Now that many of the previous Fellows have graduated and started their professional careers, our future plans involve reunions with the Fellows to discuss their current professional status, and what the REU programme has changed about how they approach their careers and in what ways it has influenced their K-12 students.

We invite researchers (including teacher researchers) to join us in debating what interdisciplinary mathematics is, and how, or if, it is any different than interdisciplinary STEM. We call on you to follow our STEM Education Department at the University of Kentucky (USA) as we model and visualize the natural world, engineer better lives for K-12 students, and study all those mathematical experiences that shape interdisciplinary STEM attitudes, notions, and beliefs. We are situating ourselves with the future interdisciplinary teacher researchers of the world.

Acknowledgements This research was supported by the National Science Foundation (NSF project # 1156848).

References

Adedokun, O. A., Bessenbacher, A. B., Parker, L. C., Kirkham, L. L., & Burgess, W. D. (2013). Research skills and STEM undergraduate research students' aspirations for research careers: Mediating effects of research self-efficacy. *Journal of Research in Science Teaching, 50*(8), 940–951.

Bodner, G. M., & Guay, R. B. (1997). The Purdue visualization of rotations test. *The Chemical Educator, 2*(4), 1–17.

Clewell, B. C., Anderson, B. T., & Thorpe, M. E. (1992). *Breaking the barriers: Helping female and minority students succeed in mathematics and science*. San Francisco: Jossey-Bass.

Committee on Science, Engineering, and Public Policy. (2006). *Rising above the gathering storm: Energizing and employing America for a brighter economic future*. Washington, D.C.: National Academies Press.

Creswell, J. W., & Plano Clark, V. L. (2007). *Designing and conducting mixed methods research*. Thousand Oaks, CA: Sage.

de Winter, J. C. F., & Dodou, D. (2010). Five-Point Likert items: t test versus Mann-Whitney-Wilcoxon, Practical Assessment. *Research and Evaluation, 15*(11).

Gatta, M., & Trigg, M. (2001). *Bridging the gap: Gender equity in science, engineering and technology*. New Brunswick, NJ: Center for Women and Work, Rutgers University.

Jacobs, V. R., Lamb, L. L. C., & Philipp, R. A. (2010). Professional noticing of children's mathematical thinking. *Journal for Research in Mathematics Education, 41,* 169–202.

Jong, C., Hodges, T., Royal, K., & Welder, R. (2015). Instruments to measure elementary preservice teachers' conceptions. *Educational Research Quarterly, 39*(1), 21–48; Social Science Premium Collection.

Kardash, C. (2000). Evaluation of an undergraduate research experience: Perceptions of undergraduate Fellows and their faculty mentors. *Journal of Educational Psychology, 92*(1), 191–201.

Kentucky STEM Task Force. (2007). *Kentucky's STEM imperative: Competing in the global economy*. Frankfort, KY: Council on Postsecondary Education STEM Task Force Report.

Landrum, R. E., & Nelsen, L. R. (2002). The undergraduate research assistantship: An analysis of the benefits. *Teaching of Psychology, 29*(1), 15–19.

Lindell, R., & Olsen, J. P. (2002). Developing the lunar phases concept inventory. *American Association of Physics Teachers Summer Meeting (Physics Education Research Conference).*

Lopatto, D. (2003). The essential features of undergraduate research. *Council on Undergraduate Research Quarterly, 23*(1), 139–142.

OECD. (2007). *PISA-2006 Science competencies for tomorrow's world, 1 (analysis)*. Paris: OECD.

OECD. (2014). *PISA-2012 Results: What students know and can do—student performance in mathematics, reading and science* (Revised Edn., Vol. I, February 2014), PISA, OECD Publishing.

OECD. (2016). *PISA-2015 Results in focus*. Paris: OECD Publishing.

Sabatini, D. A. (1997). Teaching and research synergism: The undergraduate research experience. *Journal of Professional Issues in Engineering Education & Practice, 123*(3), 98.

U.S. Department of Education. (2006). *Meeting the challenge of a changing world: Strengthening education for the 21st century*. Washington, D.C.: U.S. Government Printing Office.

U.S. Office of Science and Technology Policy. (2006). *American competitiveness initiative: Leading the world in innovation*. Washington, D.C.: U.S. Government Printing Office.

Wilhelm, J., Ganesh, B., Sherrod, S., & Ji, J. (2007). *Geometric spatial assessment*. Developed at Texas Tech University (modified in 2012).

Willis, D. A., Krueger, P. S., & Kendrick, A. (2013). The influence of a research experiences for undergraduates program on student perceptions and desire to attend graduate school. *Journal of STEM Education, 14*(2), 21–28.

Part V
Conclusion to Interdisciplinary Mathematics Education

Brian Doig

Chapter 19
Conclusion to Interdisciplinary Mathematics Education

Brian Doig and Julian Willams

Abstract We summarise the progress made by the works reported in this book for the field of Interdisciplinary Mathematics Education. It is concluded that there is still much to be done in this subfield.

Keywords Interdisciplinary · Mathematics · Education · Conclusion

In our original pre-ICME-13 assessment of the *State of the art* (Williams et al., 2016), we commented on key weaknesses in the field that research needed to address, and this book does address some of these, such as the clarification of the conceptions in the field and its theoretical bases; the different integrative and cross curricular practices studied in cases; and the policies adopted in schools breaking down disciplinary barriers.

However, we can see clearly, that our work still has a long way to go in some of these areas, e.g. the literature reviews in the field need to be based in the theoretical, and conceptual work to produce effective systematic reviews; new fit for purpose measurement tools are desperately needed if systematic quantitative work is to be conducted for evaluations that can accumulate in meta-analyses; and case studies need to be clear about the phenomena they are researching.

At repeated points in this volume interdisciplinarity has raised the question of meta disciplinary knowledge needed to work across disciplines. This was introduced previously in the *State of the art*, but only in its infancy. Disciplinary awareness becomes necessary in teacher education, in policies affording the integration of the curriculum, and in professional practices when disciplinary power has to be negotiated, and to become flexible in the greater good. But most important, we need to learn how to frame the curriculum, pedagogy and assessment for learners in ways that can encourage their own development of disciplinary awareness. It has been argued that disciplinary and interdisciplinary awareness involves knowing when *not* to use a particular discipline as much as when and how to do so: this is arguably

B. Doig
Deakin University, Burwood, Australia

J. Willams (✉)
University of Manchester, Manchester M13 9PL, UK
e-mail: Julian.williams@manchester.ac.uk

B. Doig et al. (eds.), *Interdisciplinary Mathematics Education*,
ICME-13 Monographs, https://doi.org/10.1007/978-3-030-11066-6_19

a key idea for schooling, that we might treasure the capacity of learners to become un-disciplined in this sense. But of course, learning when not to play the game also requires at least some knowledge of when and how to do so.

In all the discussions concerning IdME in the *State of the art*, and in this book, it seems to us that we have almost all missed a rather important element, one that is not particularly discussed in the literature: the distinct and special disciplinary nature of mathematics itself, and how and why this makes mathematics special to inter-disciplinary work and education. Of course, mathematics is discussed, and all the chapters and cases involve mathematics, with mathematics always implicit, and often explicit. But what is missing—or at least hardly emphasised—is a discussion of the particular features of mathematics in relation to the phenomenon of inter-disciplinarity, that is, its particular relationship with other disciplines in problem solving and inquiry, or, for example, its particular position within education, and its particular history of pedagogy and curriculum.

Actually, one thread has been mentioned, and that is the tendency for mathematics to disappear from attention in interdisciplinary problem solving, and integrated studies. The tendency for mathematics to be black-boxed or treated as a tool is part of this tendency. This is not unique to mathematics (other disciplines disappear too, unless they are deliberately upheld, and in many tasks science is irrelevant from the beginning, even when mathematics remains salient). However, interdisciplinarity may take a particular form when involving mathematics, insofar as mathematics may become regarded as a tool by *other* disciplines, as such, taking up much less cognitive space, or of being trivial in demand, compared to these other disciplines, and so less visible to the learners and less powerful than these other disciplines.

Clearly this relates to mathematics as a language of, or for, the sciences. But it also arises from the rôle of a discipline in the service of solving problems arising from outside the disciplines, e.g. modelling in 'real world problem solving', rather than for learning the discipline for its own sake. This clearly applies to all disciplines in use in problem solving, while the former is particular to the case of mathematics (and perhaps 'use of language'—if we take language use to be a discipline).

The relation between mathematics and other disciplines and sciences, then, is a particular topic that needs elucidation if we are to understand interdisciplinarity in mathematics education. Historically, the emergence of mathematics as a distinct discipline could help us: the account, in Williams and Roth (this volume), is a first attempt, and does not yet draw out the particularity of mathematics. But actually, if our purpose is to understand interdisciplinarity in the context of schools and universities, a cultural-historical framing would need to trace the emergence of: mathematics as a separate discipline; the modern school curriculum; the development of the subject teachers' professions; and the alienation of disciplines from each other, all from the interdisciplinary point of view. All this would be needed to understand how the present came to be, and what separates the school subjects, and what keeps their associated disciplines apart.

For instance, the way school subjects separate the 'interests' of teachers' professional communities is significant in explaining why the mathematics needed in some other subjects is often instrumentalised, or even trivialised. At the level of the curriculum, for instance, one sees a tendency for even Physics texts and syllabuses to find ways of reducing the necessary mathematics: topics that require minimal mathematics can be emphasised even at degree level. This can be justified by the wish to make the subject more accessible, or even more commercial. Anecdotally, some time ago, during a case study in one university engineering department, some senior staff claimed that they would be concerned about the consequences of a research team sending intending students a questionnaire about mathematics, in case this made them think mathematics was an important pre-requisite, and so decide not to apply to their university.

Then, one can add to this, the historic pedagogy of rote learning in mathematics that can be reflected in teachers of other subjects approach to dealing with mathematics in their subject teaching. In some science text books, it is not uncommon to see the mathematical relation in Physical Laws reduced to a 'cover up rule' for the Triangle with V, I, and R at its vertices (with V at the apex) thus making the mnemonic for $V = IR$ iconic. The use of mnemonics are as widespread in mathematics as in science (e.g. SOH-CAH-TOA for trigonometry, or those for the colours of the rainbow). But, if the mathematics in use in other sciences is *always* instrumentalised there, this clearly carries dangers for a 'hidden curriculum', not only for mathematics, but for the way interdisciplinary mathematics becomes understood, i.e. for the meta-knowledge of the disciplines.

A further issue is that interdisciplinary mathematics is clearly a very broad church. The range of interdisciplinary work described in the preceding chapters testifies to this, and therein lies a challenge for the future. How do we construct a unifying definition of interdisciplianary mathematics education that supports this broad range of contexts with coherence? It is reasonable to suggest that, unless a suitable, broad, definition can be created, and accepted universally, interdisciplinary mathematics education may suffer the fate of other educational waves of enthusiasm, and become a name for virtually anything that is different from the disciplinary tradition.

This lack of definition, of course, raises a further issue, and that is the lack of research into interdisciplinary mathematics education itself, to provide educators, policy-makers, and politicians, with some boundaries and facts. As mentioned earlier, the definition, and thus, the construction of a discipline, may need research that uses an Activity Theory approach, to ensure that all aspects, including dis-positions, are taken into account.

In conclusion, this volume has added to the state of the art in interdisciplinary mathematics education, but we have here posed some key questions for future research, such as "in what way is 'mathematics' special in interdisciplinary education?

References

Williams, J., Roth, W.-M., Swanson, D., Doig, B., Groves, S., Omuvwie, M., et al. (2016). *Interdisciplinary mathematics education: State of the art*. Cham: Springer. https://doi.org/10.1007/978-3-319-42267-1.

The manufacturer's authorised representative in the EU is Springer Nature Customer Service Centre GmbH, Europaplatz 3, 69115 Heidelberg, Germany. If you have any concerns regarding our products, please contact ProductSafety@springernature.com

Printed and bound by CPI Group (UK) Ltd, Croydon, CR0 4YY
05/10/2026
02228131-0008